Lecture N
Mathematics

Edited by A. Dold and B. Eckmann
Subseries: Harvard/MIT
Adviser: G. Sacks

612

Eugene M. Kleinberg

Infinitary Combinatorics and the Axiom of Determinateness

Springer-Verlag
Berlin Heidelberg New York 1977

Author
Eugene M. Kleinberg
Department of Mathematics
Massachusetts Institute
of Technology
Cambridge, MA 02139/USA

AMS Subject Classifications (1970): 02-02, 02K15, 02K35, 04-02, 04A20, 05A17

ISBN 3-540-08440-1 Springer-Verlag Berlin Heidelberg New York
ISBN 0-387-08440-1 Springer-Verlag New York Heidelberg Berlin

Printing and binding: Beltz Offsetdruck, Hemsbach/Bergstr.
2141/3140-543210

For Evie

I know a reasonable woman,
Handsome and witty, yet a friend.
Not warped by passion, awed by rumor,
Not grave through pride,
                    or gay through folly,
An equal mixture of good humor,
And sensible soft melancholy.

— Alexander Pope

Acknowledgements

These notes are distilled from a number of courses I taught at M.I.T. between 1970 and 1976. They first appeared in draft form as material for a graduate course given here last fall.

The subject matter itself was initiated mainly through the work of Martin and Solovay in the late 1960's. I owe a strong intellectual debt to both of them.

Special credit is due to Professor J. M. Henle. He is a major contributor to the development of the field, and many of his results appear throughout these notes.

In preparing this manuscript, I had extensive and expert help from many of my students, most notably from Arthur Apter.

Finally, I owe special thanks to Denise Borsuk for her patience, skill, and intuition in transforming my assorted notes to type.

Cambridge, Mass.
22 April 1977

Table of Contents

Infinitary Combinatorics
and the
Axiom of Determinateness

Introduction

Although it continued to draw heavily on ideas and notions discovered many years earlier, set theory entered its modern era in the early 1960's on the heels of Cohen's discovery of the method of forcing and Scott's discovery of the relationship between large cardinal axioms and constructible sets.

At the time, Cohen's results inspired the greater flurry of activity. On the one hand, a long unsolved problem of great interest had been settled,and on the other, the proof which was used immediately presented set theorists with a powerful tool for establishing relative consistency results. Over the years, however, it has turned out to be Scott's theorem rather than Cohen's which inspired the deeper and more exciting work.

Scott had shown that so called large cardinal axioms (in his case, the existence of a measurable cardinal) could entail seemingly unrelated consequences about the size of the universe of sets. Thus, whereas Cohen's result told us that the current axioms of Zermelo and Fraenkel were insufficient to resolve a number of fundamental mathematical questions, Scott's theorem gave us hope for their ultimate resolution through the use of new axioms. (Quite remarkably, Gödel had prophesized results such as this in a 1946 address given at Princeton.)

A few years after Scott, Rowbottom, in a truly beautiful and original piece of work, showed that a large cardinal axiom weaker than that used by Scott yielded the existence of nonconstructible sets of integers. This was extremely remarkable, for it showed that an assumption about the existence of an extremely large cardinal having certain properties could yield a new result about a set so small as a set of integers.

In the years following, new axioms for set theory, their relationships with one another, and their consequences for set theory in general, were studied heavily. Powerful structure theorems for the universe of constructible sets were proved by Silver using so called Erdös cardinal axioms, Kunen developed refined techniques using iterated ultrapowers, and notions such as measurable cardinals, Ramsey cardinals, Rowbottom cardinals, and Jonsson cardinals became standard set theoretic fare.

o

There was another axiom being extensively studied at this time. It was referred to as the axiom of determinateness and was an axiom different from the others in more ways than one. First considered by Ulam, the mathematician who also discovered the notion of measurable cardinal, the axiom of determinateness was not an axiom of the form of those mentioned earlier, "there exists a set x with the following property: ......", but rather was of the form "the set of real numbers has the following property: ...... ". The axiom had another bizarre feature. It implied the non-existence of a well-ordering of the reals, and as such it was in conflict with the axiom of choice.

This is no place for a detailed philosophical discussion about the virtues of one axiom over another, but the special nature of the axiom of determinateness deserves some discussion. Why were set theorists so drawn to study this axiom, drawn, in fact, to the point where it became the key area of research for all but a few of the best in the field? The answer is not simple and would best be given after one had read somewhat into the subject. A few points, however, should be made here:

1. Contradicting the axiom of choice is not, in and of itself, grounds for rejecting the axiom of determinateness. Set theorists long ago segregated the axiom of choice from the standard axioms of set theory due to its somewhat tenuous nature. Mathematicians in general had been aware of problems with the axiom of choice, and following the discovery of such things as the Banach-Tarski paradox, analysts avoided it whenever possible.
2. Very rarely is the full power of a given axiom used in proving theorems. Mathematicians often work in the atmosphere of conceptual simplicity presented by a strong axiom, and following their discovery of the desired theorem, they go back to the proof and examine precisely how many special assumptions were _really_ needed. Often, one is able to recast the proof so that no special new axiom is needed or so that only a weaker one is. In this way, you have your desired theorem by a means you never would have discovered directly.

The axiom of determinateness has turned out to have an extraordinary impact on modern set theory. It has presented a powerful approach to certain questions, an approach which has been used with remarkable success in a whole array of results ranging from those requiring only weak (and sometimes outright provable) instances of the axiom to those which seem to rely heavily on strong instances.

———— o ————

In what follows, we shall concern ourselves with the full axiom of determinateness (henceforth to be known as AD), and in particular, shall consider the consequences of AD in the field of combinatorial set theory. The history here is as follows: in about 1966, Solovay proved that AD implied the existence of measurable cardinals. This result immediately caused wide interest for a number of reasons. First off, although AD had earlier yielded remarkable facts about sets of reals (eg: AD implies that all sets of reals are Lebesgue measurable (Mycielski and Swierczkowski)), the very nature of the axiom seemed to defy its proving anything about cardinals themselves. Furthermore, that AD should yield one of the key large cardinal hypotheses of the day was most interesting in itself, but on top of this, the cardinal shown to be measurable was $\aleph_1$, and this was nothing short of extraordinary. The proof used for this result was quite lovely, using a blend of both set theory and recursion theory, and although it hinted of further promise for AD as an axiom of interest, it was not for a couple of years that new and striking results were to appear. In 1968, Martin found a new proof of the measurability of $\aleph_1$ from AD. His ideas were simple and elegant yet of such power that within a couple of weeks, Solovay had cleverly modified the argument to show that under the assumption of AD, $\aleph_2$ is a measurable cardinal. The trend was now set, and everybody's goal became to show each $\aleph_n$ measurable from AD. In the great effort that followed, a good deal of machinery was constructed using intricate definability/recursion theoretic considerations, yet the goal remained untouched. The reason for such striking nonsuccess is, in retrospect, clear —— the goal was false. In about 1970, drawing upon an enormous amount of the previous two years work with the axiom, Martin showed AD to imply that each $\aleph_n$, for $n > 2$, was singular with cofinality $\aleph_2$. To say the least, this bizarre result sat on an obscure proof which was never universally understood, and subsequent work with AD passed over the $\aleph_n$ in favor of larger, more easily understood, cardinals. Even this project, however, was abandoned within a few years as it became increasingly technical and cumbersome.

The situation changed with the introduction of new and elegant methods in 1975. The theorems of Solovay and Martin can now be reproved as a natural progression of results so that the entire sequence of theorems holds together in one unified piece. Furthermore, these new proofs are basically combinatorial, are easier to understand, and make clear why $\aleph_1$ and $\aleph_2$ should be measurable yet $\aleph_3$, $\aleph_4$, $\aleph_5$,... singular with cofinality $\aleph_2$. And most surprisingly, these new methods go on to prove that despite their being singular, each $\aleph_n$, for $n > 2$, is a Jonsson cardinal, and $\aleph_\omega$, their limit, is a Rowbottom cardinal. So not only does AD imply that the first two uncountable cardinals satisfy well-known large cardinal properties, the following $\omega + 1$ — many do as well.

The means toward these new methods and consequences of AD grew from the notion of partition relation, the single most important concept in modern large cardinal theory. Its origins were based in the late 1940's when Erdös and his school began looking at generalizations of Ramsey's remarkable theorem of 1928. From these generalizations eventually grew the notions of Rowbottom cardinal (the means for the initial improvement of Scott's theorem), Erdös cardinal (the means for Silver's ultimate improvement) and Jonsson cardinal (the weakest cardinal Kunen was finally able to derive Silver's results from).

A further generalization, that of the infinite exponent partition relation, was largely ignored for many years, for like the axiom of determinateness, infinite exponent partition relations imply the nonexistence of well-orderings of certain sets. In 1968, however, soon after AD had shown $\aleph_2$ to be measurable, it was proved that any cardinal satisfying an appropriate infinite exponent partition relation is measurable. The production of measurable cardinals from such relations turned out to be easier than the production of measurable cardinals from AD, and furthermore, with partition relations, one had a complete characterization of all normal measures produced. At the time, however, AD was the established axiom, and people had little feel for the nature or strength of infinite exponent partition relations. They were thus treated as a curiosity until late 1968 when Martin proved from AD that $\aleph_1$ satisfied infinite exponent partition relations. Further results along these lines followed, and by 1973 the stage had been set. Martin had shown $\aleph_1$ to satisfy $\aleph_1 \longrightarrow (\aleph_1)^{\aleph_1}$, the strongest possible infinite exponent partition relation.

Our unified derivation of AD's consequences for cardinals through $\aleph_\omega$ makes use of the partition relation $\aleph_1 \longrightarrow (\aleph_1)^{\aleph_1}$

and does not really depend upon AD at all.

In what follows, the theory of infinite exponent partition relations is studied in detail. Along the way, we will prove that any κ satisfying $\kappa \longrightarrow (\kappa)^{\kappa}$ is measurable, has a measurable κ' larger than it, and has countably many singular Jonsson cardinals larger than κ' whose limit is a Rowbottom cardinal. This purely combinatorial theorem can stand by itself, or it can be joined with the well-known results of Martin and Solovay on AD's consequences for $\aleph_1$ (covered in Chapter II of these notes) to yield the full sequence of consequences of AD mentioned earlier.

Unlike most combinatorial results whose proofs begin each time from scratch, those results covered in later chapters of these notes depend heavily on results covered earlier.

Aside from Chapter II which derives various properties of $\aleph_1$ from AD, all other chapters are primarily combinatorial. One needs Chapter I to understand Chapter III, III to understand IV, and so forth through Chapter VI. Chapter VII can be read using only Chapter I.

Finally, the notes have been designed so that one who does not read deeply into the later chapters would still benefit from earlier chapters. For example, by the end of Chapter IV one would have a complete proof that assuming AD, $\aleph_1$ and $\aleph_2$ are measurable and $\aleph_3$ is a singular Jonsson cardinal cofinal with $\aleph_2$. Furthermore, if the reader is uncomfortable with noncombinatorial set theory, he can skip Chapter II entirely. This is because Chapter II is only concerned with the derivation, from AD, of two combinatorial facts,

$$\aleph_1 \longrightarrow (\aleph_1)^{\aleph_1} \quad \text{and}$$

$$\aleph_1^{\aleph_1}/\mu_\omega = \aleph_2 \quad \text{—}$$

all further work is completely combinatorial and builds solely on these two combinatorial facts.

Chapter I

Let γ be any uncountable regular cardinal. A subset X of γ is said to be <u>ω-closed</u> if the sup of any increasing ω-sequence of elements of X is a member of X. In the summer of 1968, H. Friedman asked the following question:

> Is it consistent with ZFC for there to be a regular uncountable cardinal γ such that the collection of those subsets of γ which contain ω-closed unbounded sets is an ultrafilter?

The unique feature to this question was, of course, Friedman's use of "ω-closed" rather than just "closed" — it was well-known that the axiom of choice ruled out such consistency otherwise.

As things turned out, however, the answer was still no. Here is the reasoning:

<u>Lemma 1.1</u>: For any regular uncountable cardinal γ, the intersection of any fewer than γ-many ω-closed unbounded subsets of γ is itself ω-closed and unbounded.

<u>Proof</u>: Let $\{B_\alpha | \alpha < \lambda < \gamma\}$ be a given collection of ω-closed unbounded subsets of γ. Clearly, $\bigcap_{\alpha<\lambda} B_\alpha$ is ω-closed. To see that it is unbounded, we proceed as follows: suppose $\nu < \gamma$. Let α_0 be the least member of B_0 larger than ν, let α_1 be the least member of B_1 larger than α_0, and, in general, let α_η be the least member of B_η larger than every α_τ for $\tau < \eta$. Since γ is regular and $\lambda < \gamma$,

$\bigcup_{\eta<\lambda} \alpha_\eta < \gamma$, and so we can repeat the process: let α_0' be the least member of B_0 larger than $\bigcup_{\eta<\lambda} \alpha_\eta$, α_1' the least member of B_1 larger than α_0', and, in general, α_η' the least member of B_η larger than every α_τ' for $\tau < \eta$. We continue ad infinitum in this way, and eventually produce increasing ω-sequences

$$\{\alpha_0, \alpha_0', \alpha_0'', \alpha_0''', \alpha_0'''', \ldots\} \subseteq B_0$$

$$\{\alpha_1, \alpha_1', \alpha_1'', \alpha_1''', \alpha_1'''', \ldots\} \subseteq B_1$$

$$\{\alpha_2, \alpha_2', \alpha_2'', \alpha_2''', \alpha_2'''', \ldots\} \subseteq B_2$$

As each B_η is ω-closed, $\bigcup\{\alpha_\eta, \alpha_\eta', \alpha_\eta'', \alpha_\eta''', \ldots\} \in B_\eta$ for each $\eta < \lambda$. But we constructed our sequences in such a way that <u>they all have the same sup</u>. This sup is thus a member of $\bigcap_{\eta<\eta} B_\eta$ larger than ν. Since ν was arbitrary, $\bigcap_{\eta<\lambda} B_\eta$ must be unbounded. □

<u>Definition</u>: For any regular uncountable cardinal γ, let W_γ denote the collection of those subsets of γ which contain ω-closed unbounded sets.

<u>Lemma 1.2</u>: Let γ be any regular uncountable cardinal. Then assuming the axiom of choice, W_γ is a γ-additive filter on γ.

Proof: This follows quite easily from 1.1. Clearly, $\phi \notin W_\gamma$, and, equally clearly, if $A \in W_\gamma$ and $B \supseteq A$, then $B \in W_\gamma$. To see that W_γ is γ-additive, suppose $\{A_\eta | \eta < \lambda < \gamma\}$ is a given collection of members of W_γ. Using the axiom of choice, let us choose an ω-closed unbounded subset D_η of A_η for each $\eta < \lambda$. By 1.1, $\bigcap_{\eta<\lambda} D_\eta$ is an ω-closed unbounded set, and as $\bigcap_{\eta<\lambda} D_\eta \subseteq \bigcap_{\eta<\lambda} A_\eta$, this tells us that $\bigcap_{\eta<\lambda} A_\eta \in W_\gamma$. □

Theorem 1.3: Assuming the axiom of choice, W_γ is not an ultrafilter for any regular uncountable cardinal γ.

Proof: Assume γ is the least uncountable regular cardinal such that W_γ is an ultrafilter on γ. Then the following Π^1_2 sentence would be true in the structure $\langle R(\gamma), \varepsilon \rangle$:

(*) $$\forall^1 X \exists^1 Y \big((Y \subseteq X \vee Y \subseteq X^c) \wedge \text{"}Y \text{ is } \omega\text{-closed and unbounded"} \big)$$

By 1.2, the axiom of choice implies that γ is a measurable cardinal, and since (again by the axiom of choice) any measurable cardinal is Π^2_1-indescribable, the sentence (*) must be true in $\langle R(\alpha), \varepsilon \rangle$ for some α less than γ. Since this would imply that W_α was an ultrafilter on α, we have a contradiction to our leastness assumption on γ. The theorem now follows. □

Admittedly, our proof of theorem 1.3 is somewhat obscure. Here is a more direct combinatorial argument:

Lemma 1.4: Assume γ is a regular uncountable cardinal and that W_γ is an ultrafilter. Then assuming the axiom of choice, the measure μ on γ defined by "$\mu(A) = 1$ if $A \in W_\gamma$" is a normal γ-additive measure on γ and $\mu(\{\alpha |$ the cofinality of α is $\omega\}) = 1$.

Proof: It is immediate from 1.2 that μ is a γ-additive measure on γ. Furthermore, since $\{\alpha \in \gamma |$ the cofinality of α is $\omega\}$ is ω-closed and unbounded, it clearly has μ-measure 1. Thus, we need only show that μ is normal. There are two ways to do this. A more conventional argument would be to prove via a method reminiscent of that used in proving 1.1 and 1.2 that the diagonal intersection of any κ-many sets of μ-measure 1 has μ-measure 1. Here, however, is another proof which requires only countably many choices: assume that the measure μ is not normal and that $g:\gamma \longrightarrow \gamma$ is such that $\mu(\{\alpha | g(\alpha) < \alpha\}) = 1$, yet $\mu(\{\alpha | g(\alpha) = \alpha_0\}) = 0$ for each $\alpha_0 < \gamma$. Let $A =_{df} \{\alpha | g(\alpha) < \alpha\}$. Now μ is γ-additive, and the fact that g is not constant almost everywhere implies that for every $\eta < \gamma$, $\mu(g^{-1}\eta) = 0$. We can thus define an ω-sequence of ordinals and sets of ordinals as follows: $\beta_0 =_{df} 0$, A_0 is an ω-closed unbounded subset of A satisfying "$\alpha \in A_0 \Longrightarrow g(\alpha) > 0$", and α_0 is the least member of A_0. Proceeding inductively, once β_n, A_n and α_n have been defined, we define β_{n+1} to be the least member of the range of g on A_n larger than α_n, A_{n+1} to be an ω-closed unbounded subset of A_n

satisfying "$\alpha \in A_{n+1} \Longrightarrow g(\alpha) > \beta_{n+1}$", and α_{n+1} to be the least member of A_{n+1}. Now since, for every n, a final segment of the sequence $\alpha_1, \alpha_2, \alpha_3, \alpha_4, \alpha_5, \ldots\ldots$ lies entirely within the ω-closed unbounded set A_n, the sup of that sequence, $\bigcup_{n<\omega} \alpha_n$, must be a member of $\bigcap_{n<\omega} A_n$. This immediately implies that $g(\bigcup_{n<\omega} \alpha_n) \geq \bigcup_{n<\omega} \beta_n$. Since $\bigcup_{n<\omega} \alpha_n = \bigcup_{n<\omega} \beta_n$, we then have $g(\bigcup_{n<\omega} \alpha_n) \geq \bigcup_{n<\omega} \alpha_n$. But $\bigcap_{n<\omega} A_n \subseteq A$, which implies that $g(\bigcup_{n<\omega} \alpha_n) < \bigcup_{n<\omega} \alpha_n$. This contradiction yields our result.

□

Lemma 1.5: Assume the axiom of choice. Then given any normal measure ν on a measurable cardinal κ,

$$\nu(\{\alpha \mid \alpha \text{ is a regular cardinal}\}) = 1.$$

Proof: Suppose $\nu(\{\alpha < \kappa \mid \alpha \text{ is singular}\}) = 1$. We will derive a contradiction: Let $g : \kappa \longrightarrow \kappa$ be given by $g(\alpha)$ = "the cofinality of α". As almost every α less than κ is singular, $g(\alpha) < \alpha$ for almost every α, and so, as ν is normal, there exists an α_0 less than κ such that $g(\alpha) = \alpha_0$ for almost every α. Let $A = \{\alpha < \kappa \mid g(\alpha) = \alpha_0\}$. Now for each α in A, let us use the axiom of choice to choose an α_0-sequence of ordinals with sup α. Let us denote the sequence associated with an ordinal α as $\alpha^0, \alpha^1, \alpha^2, \ldots, \alpha^\eta, \ldots (\eta < \alpha_0)$. Then for each $\eta < \alpha_0$, the function $g_\eta : A \longrightarrow \kappa$ given by $g_\eta(\alpha) =_{df} \alpha^\eta$ satisfies

$\nu(\{\alpha \mid g_\eta(\alpha) < \alpha\}) = 1$. Thus by normality there are, for each $\eta < \alpha_0$, sets of ν-measure 1 A_η and ordinals β_η such that $g_\eta'' A_\eta = \{\beta_\eta\}$. Since $\bigcap_{\eta<\alpha_0} A_\eta$ has ν-measure 1, it contains (at least) two distinct elements, say $\alpha_1 < \alpha_2$. Our proof is now almost complete: Since α_1 and α_2 are in $\bigcap_{\eta<\alpha_0} A_\eta$, $g_\eta(\alpha_1) = g_\eta(\alpha_2)$ for every $\eta < \alpha_0$, and as $g_0(\alpha_1), g_1(\alpha_1), g_2(\alpha_1), g_3(\alpha_1), \ldots\ldots, g_\eta(\alpha_1), \ldots\ldots$ is an α_0-sequence with sup α_1 and $g_0(\alpha_2), g_1(\alpha_2), g_2(\alpha_2), g_3(\alpha_2), \ldots\ldots, g_\eta(\alpha_2), \ldots\ldots$ an α_0-sequence with sup α_2, we would have $\alpha_1 = \alpha_2$, an obvious contradiction. The lemma is thus proved.

□

Clearly, Lemmas 1.4 and 1.5 contradict one another, and as such comprise a proof of Theorem 1.3.

——————— o ———————

Now what happens if we don't use the axiom of choice? Is it then possible for W_γ to be an ultrafilter for some γ? This question marks the beginning of our work.

<u>Definition:</u> Let x be a given set of ordinals and let α be a given ordinal. Then we denote by $[x]^\alpha$ the collection of α-sequences from x, that is, the collection of all those subsets of x of order-type α.

Definition: Let α, β and γ be ordinals, $\alpha \leq \gamma$. Then $\gamma \longrightarrow (\gamma)^{\alpha}_{\beta}$ denotes the following partition relation: given any partition $F: [\gamma]^{\alpha} \longrightarrow \beta$ of $[\gamma]^{\alpha}$ into β-many pieces, there exists a subset C of γ of cardinality γ homogeneous for the partition F in the sense that $\overline{\overline{F''[C]^{\alpha}}} = 1$.

Clearly, $\gamma \longrightarrow (\gamma)^{\alpha}_{\beta}$ is false if $\beta \geq \gamma$. It is well known that $\gamma \longrightarrow (\gamma)^{2}_{2}$ is true for $\gamma = \omega$ (this is just Ramsey's theorem), and that if $\gamma \longrightarrow (\gamma)^{2}_{2}$ is true for any ordinal γ other than ω, then that γ must be a strongly inaccessible cardinal (a theorem of Erdös). (By convention, we shall delete the subscript β in $\gamma \longrightarrow (\gamma)^{\alpha}_{\beta}$ whenever $\beta = 2$. Thus $\gamma \longrightarrow (\gamma)^{2}$ is the same as $\gamma \longrightarrow (\gamma)^{2}_{2}$.)

Partition relations of the form $\gamma \longrightarrow (\gamma)^{\alpha}_{\beta}$ where $\alpha \geq \omega$ are known as infinite exponent partition relations. They are our fundamental concern.

Lemma 1.6: Assume that γ satisfies the relation $\gamma \longrightarrow (\gamma)^{\omega}$. Then W_{γ} is an ultrafilter.

Proof: This result is really quite simple modulo one technical fact:

Definition: If x is a given set of ordinals, let us denote by $(x)_{\omega}$ the collection of sups of all ω-sequences from x, that is, $(x)_{\omega} =_{dn} \{\bigcup p \mid p \in [x]^{\omega}\}$.

Technical Fact: If x is any unbounded subset of a regular uncountable cardinal γ, then $(x)_\omega$ is an ω-closed unbounded subset of γ.

(Proof of Technical Fact: It is easy to see that $(x)_\omega$ is unbounded in γ, for if $\nu < \gamma$, $\overline{\overline{x - \nu}} = \gamma$, and so $[x - \nu]^\omega$ is nonempty. If $p \in [x - \nu]^\omega$, then $\bigcup p \in (x)_\omega$ and $\bigcup p > \nu$. As ν was arbitrary, $(x)_\omega$ must be unbounded. Now to see that $(x)_\omega$ is ω-closed, suppose $p \in [(x)_\omega]^\omega$. We must show that $\bigcup p \in (x)_\omega$. For each $n < \omega$, let p_n denote the nth largest ordinal in the sequence p. As p_{n+1} is a limit of an ω-sequence from x and $p_{n+1} > p_n$ for each $n < \omega$, let, for each $n < \omega$, β_n be the least member of x such that $p_n < \beta_n < p_{n+1}$. Then $\beta_0, \beta_1, \beta_2, \beta_3, \beta_4, \beta_5, \ldots\ldots$ is an ω-sequence of elements of x with limit equal to $\bigcup p$. Thus $\bigcup p = \bigcup\{\beta_n | n < \omega\} \in (x)_\omega$. ☒)

Proof of Lemma 1.6: Suppose A is a given subset of γ. We must show that either A or A^c contains an ω-closed unbounded subset. Let the partition $G: [\gamma]^\omega \longrightarrow 2$ be given by

$$G(p) = \begin{cases} 0 & \text{if} \quad \bigcup p \in A \\ 1 & \text{if} \quad \bigcup p \in A^c, \end{cases}$$

and, by using $\gamma \longrightarrow (\gamma)^\omega$, let C be a size γ subset of γ homogeneous for G (i.e., $\overline{\overline{G''[C]^\omega}} = 1$). By definition of G, (and our technical fact), if $G''[C]^{\omega} = \{0\}$, then $(C)_{\omega}$ is an ω-closed unbounded subset of A, whereas if $G''[C]^{\omega} = \{1\}$, it is an ω-closed unbounded subset of A^{c}. In either case, we have our result.

□

This surprising result tells us, then, that under certain assumptions, W_{γ} can be an ultrafilter.

Now since we are working in set theory without the axiom of choice, W_{γ} being an ultrafilter on γ does not necessarily imply that γ is a measurable cardinal. Lemma 1.2 simply cannot be applied without the axiom of choice, and so we do not know, in our present context, anything about the additivity of the filter W_{γ}. Partition relations, however, come to our aid once again:

Lemma 1.7: Assume that γ satisfies the partition relation $\gamma \longrightarrow (\gamma)^{\omega}_{\lambda}$. Then the intersection of any λ-many members of W_{γ} is a member of W_{γ}, i.e., W_{γ} is λ^{+}-additive.

Proof: Suppose we are given sets A_{α} for $0 < \alpha < \lambda$ in W_{γ}. We wish to show that $\bigcap_{\alpha<\lambda} A_{\alpha}$ is in W_{γ}, and so must prove that $\bigcap_{\alpha<\lambda} A_{\alpha}$ contains an ω-closed unbounded subset of γ. Let the partition $G: [\gamma]^{\omega} \longrightarrow \lambda$ be given by

$$G(p) = \begin{cases} \text{the least } \alpha \text{ such that } \\ Up \notin A_\alpha \quad \text{if} \quad Up \notin \bigcap_{\alpha<\lambda} A_\alpha \\ \\ 0 \quad \text{otherwise,} \end{cases}$$

and let C be a size γ subset of γ homogeneous for G. It is easy to see that we in fact have $G''[C]^\omega = \{0\}$, for if $G''[C]^\omega = \{\alpha\}$ for some $\alpha > 0$, we would have that $(C)_\omega$ was an ω-closed unbounded set disjoint from A_α. Since A_α itself contains an ω-closed unbounded subset, this would contradict lemma 1.1. We thus have that $G''[C]^\omega = \{0\}$, and by the definition of G, $(C)_\omega \subseteq \bigcap_{\alpha<\lambda} A_\alpha$. Since $(C)_\omega$ is ω-closed and unbounded, we have our proof. □

From lemmas 1.6 and 1.7, we immediately have the following result.

<u>Theorem 1.8</u>: Assume that γ satisfies $\gamma \longrightarrow (\gamma)^\omega_\lambda$ for every λ less than γ. Then γ is a measurable cardinal.

It is important to note that our proof of theorem 1.8 yields more than just the measurability of γ — it shows that the filter generated by the collection of ω-closed unbounded sets, W_γ, is itself a γ-additive measure[(1)] on γ. This is, actually, a special occurence. For example, Martin and Mitchell have shown that if it is consistent for W_κ to be a measure on a cardinal κ, then it is consistent for there to exist many measurable cardinals. (See fact A at the end of Chapter VI.)

Given that W_γ is a γ-additive ultrafilter on a given cardinal γ, what can we say about the existence of normal measures on γ? Is W_γ itself normal? Keep in mind that lemma 1.4 does not apply as we may not use the axiom of choice.

As is well-known, normal measures are used strongly in most proofs of consequences of measurable cardinals, and so we would like to know that we have normal measures in our current context. Yet in any of the standard constructions of normal measures, a certain amount of the axiom of choice is used. Once again, we get our desired result with an appeal to an infinite exponent partition relation:

Lemma 1.9: Assume W_γ is a γ-additive filter on γ. Then $\gamma \longrightarrow (\gamma)^\omega$ implies that the measure corresponding to W_γ is normal.

Proof: Let μ denote the measure on γ associated with W_γ. Suppose $g : \gamma \longrightarrow \gamma$ is a given map such that $\mu(\{\alpha | g(\alpha) < \alpha\}) = 1$. We wish to find a $\beta_0 < \gamma$ such that

(1) We will freely identify ultrafilters on cardinals and measures on cardinals, a given ultrafilter being simply the collection of sets of measure 1.

$\mu(\{\alpha|g(\alpha) = \beta_0\}) = 1$. Let A be an ω-closed unbounded subset of $\{\alpha|g(\alpha) < \alpha\}$. Then since A can be put into 1-1 correspondence with all of γ, the partition relation $\gamma \longrightarrow (\gamma)^\omega$ tells us that, in fact, any partition of the set of ω-sequences from A has a size γ homogeneous set. Keeping this in mind, let $G : [A]^\omega \longrightarrow 2$ be given by

$$G(p) = 0 \quad \text{iff} \quad \text{"}g(\bigcup p) \text{ is less than the least member of } p\text{"},$$

and let C be a size γ subset of γ homogeneous for G.

Claim: $G''[C]^\omega = \{0\}$. (Proof of claim: Suppose $p \in [C]^\omega$. Since $C \subseteq A$ and A is ω-closed, $\bigcup p \in A$, and so $g(\bigcup p) < \bigcup p$. Let p' be that ω-subsequence of p consisting of the members of p larger than $g(\bigcup p)$. Then $p' \in [C]^\omega$, $\bigcup p' = \bigcup p$, and yet $g(\bigcup p')$ is less than the least member of p'. Thus $G(p') = 0$, and since C is homogeneous, $G''[C]^\omega = \{0\}$. $\boxtimes$)

Now let α_0 denote the least member of C.

Claim: $g''(C)_\omega \subseteq \alpha_0$. (Proof of claim: Given any $p \in [C]^\omega$, $\{\alpha_0\}\bigcup p \in [C]^\omega$, and so by the previous claim, $G(\{\alpha_0\}\bigcup p) = 0$. By the definition of G, this implies $g(\bigcup p) < \alpha_0$, and as p was arbitrary, $g''(C)_\omega \subseteq \alpha_0$. $\boxtimes$)

Since $g''(C)_\omega \subseteq \alpha_0$, we must have $(C)_\omega \subseteq \bigcup_{\beta<\alpha_0} g^{-1}\{\beta\}$. As μ is a γ-additive measure on γ, the fact that $\mu((C)_\omega) = 1$ tells us that $\mu(g^{-1}\{\beta_0\}) = 1$ for some $\beta_0 < \alpha_0$. Thus $\mu(\{\alpha|g(\alpha) = \beta_0\}) = 1$. $\square$

We can summarize the work of this section in the following result:

Theorem 1.10: Assume that γ satisfies the partition relation $\gamma \longrightarrow (\gamma)^{\omega}_{\lambda}$ for each $\lambda < \gamma$. Then γ is a measurable cardinal, and the function μ from 2^{γ} into $\{0,1\}$ given by "$\mu(A) = 1$ iff A contains an ω-closed unbounded set" is a nontrivial γ-additive normal measure on γ.

Proof: Immediate from lemmas 1.6, 1.7, and 1.9.

These results show just the tip of the iceberg. Infinite exponent partition relations have extraordinary power as will become clear as our work proceeds. In the next section, we will derive such relations from the axiom of determinateness, but it will ultimately be the partition relations themselves which give simple and elegant proofs of AD's known consequences and go on to establish new consequences.

Chapter II

The axiom of determinacy can best be described in terms of two person games of infinite length. Given a subset A of ${}^{\omega}\omega$ we consider a game G_A defined as follows:

> G_A is played between two players, I and II. Initially I begins play by writing a member n_1 of ω, II responds by writing n_2, I responds with n_3, and so forth ad inf. In this way, I and II together produce an ω-sequence of members of ω, $n_1, n_2, n_3, n_4, n_5, \ldots,$ and the payoff is that I <u>wins</u> this play of the game G_A iff this ω-sequence is a member of A.

A <u>strategy</u> for such a game G_A is simply a function from the set of finite sequences of natural numbers into the set of natural numbers, the strategy f for a given player telling him to write $f(\langle n_1, n_2, \ldots, n_i\rangle)$ as his next move after the initial play $n_1, n_2, \ldots, n_i$ in any game.

A strategy for a given game G_A is said to be a <u>winning strategy</u> for a given player (I or II) if that player using that strategy wins every play of G_A.

A subset A of ${}^{\omega}\omega$ is said to be <u>determinate</u> if one of the players has a winning strategy for G_A.

The <u>axiom of determinacy</u> (AD) is simply the assertion that every subset of ${}^{\omega}\omega$ is determinate.

———— o ————

The conflict between AD and the existence of a well-ordering of the reals is almost immediate — one can simply diagonalize (by induction) over any well-ordered sequence of strategies to produce a set of reals A for which none of those strategies is winning in G_A.

In the early 1960's, Mycielski and his school began relating AD to classical notions from mathematics, and this marked the beginning of heightened interest in the axiom. Here is a simple result from that period:

Theorem 2.1 (Mycielski, Swierczkowski): Assuming AD, every set of reals is Lebesgue measurable.

Proof (Due to L. Harrington): Without loss of generality, we consider sets of reals contained in the closed interval $[0,1]$. Using the notion of infinitary expansion (see [10]), there is a unique 1-1 correspondence between real numbers in $[0,1]$ and members of ${}^{\omega}\omega$. Viewing reals in this way, as infinite sequences of natural numbers, we can proceed as follows: it is routine to see that in order to show every set of reals Lebesgue measurable, it suffices to show that any set of reals with inner measure 0 has outer measure 0. Suppose, then, that B is a set with inner measure 0, and let $\varepsilon > 0$ be given. We wish to find an open covering of B having measure $\leq \varepsilon$. Let us assume that we have systematically assigned to each finite union of open intervals with rational endpoints a unique nonnegative integer. Then using this coding of intervals in terms of numbers, let us consider the infinite game whose payoff is as follows: player II wins the play $n_1, n_2, n_3, n_4, n_5, \ldots$ iff both

(1) his moves, $n_2, n_4, n_6, n_8, n_{10}, \ldots,$ are the code numbers for a sequence of unions of intervals such that for each i, the union of intervals coded by n_{2i} has measure less than $\varepsilon/2^{4i}$, and

(2) if player I's moves, $n_1, n_3, n_5, n_7, n_9, \ldots,$ when written $.n_1 n_3 n_5 n_7 n_9 \ldots$ constitute a binary expansion of a member of B, then that member of B lies in one of the unions of intervals coded by $n_2, n_4, n_6, n_8, n_{10}, \ldots$.

Now despite the fact that we have described this game using words from English language, it is routine to see that for some set A contained in ${}^\omega\omega$, I and II are just playing G_A. Thus, by the hypothesis of our theorem, let f be a winning strategy for this game.

If we can prove that f is actually a winning strategy <u>for player II</u>, we would have our desired open cover of B. For in this case, by the definition of the game, one can easily check that if X denotes the collection of all those unions of intervals whose code numbers appear as f is played against all possible sequences of 0's and 1's as I's moves, then X constitutes an open covering of B of measure at most ε. Thus to complete our proof, we need only show that player I can never have a winning strategy for the game described above. This is fairly simple. For suppose g were a winning strategy for player I for the above game. Then we can turn g into a map g^* from $[0,1]$ into $[0,1]$ as follows: given a real r in $[0,1]$ let $n_2, n_4, n_6, n_8, n_{10}, \ldots$, be r's unique infinitary expansion. Then $g^*(r)$ is defined to be the real with binary expansion $.n_1 n_3 n_5 n_7 n_9, \ldots$ where $n_1, n_3, n_5, n_7, n_9, \ldots$ is such that $n_1, n_2, n_3, n_4, n_5, n_6, n_7, n_8, n_9, n_{10}, \ldots$ is the play of G_A where g is used against player II playing $n_2, n_4, n_6, n_8, n_{10}, \ldots$. It is routine to see that g^* is continuous at each irrational, and hence, if we denote by P the range of g^*, P is an analytic set. Furthermore, since g is assumed to be a winning strategy for player I, P is a subset of B. As is well-known, analytic sets are Lebesgue measurable, and so, as the inner measure of B is 0, P must have measure 0. Thus, let $n_2, n_4, n_6, n_8, n_{10}, \ldots$ be a sequence of integers such that the unions of intervals coded by them form an open covering of P and such that for any i, the union of intervals coded by n_{2i} has measure at most $\varepsilon/2^{4i}$. If we now consider a play of our game where II simply plays the integers $n_2, n_4, n_6, n_8, n_{10}, \ldots$ against I's playing via the strategy g, the resulting play is a win for II. This contradicts the assumption that g is a winning strategy for I. The theorem follows.

□

We will center our concern not with classical consequences of AD, but rather with modern set theoretic ones. For the remainder of this chapter, we will work in Zermelo-Fraenkel set theory with the axiom of dependent choice.

For the sake of motivation, let us look at the problem of trying to prove that the filter of sets containing ω-closed unbounded subsets of $\aleph_1$ is an ultrafilter. In an earlier chapter, we learned how to do this using the partition relation $\aleph_1 \longrightarrow (\aleph_1)^{\omega}$ — here we will use games and strategies:

> suppose that Q is a given subset of $\aleph_1$. We wish to prove that either Q or Q^c contains an ω-closed unbounded set, and so we consider the following game played between two players I and II: I and II move alternately (beginning with I), writing at each move an <u>ordinal less than $\aleph_1$</u>. If either player fails to write an ordinal larger than that just written by his opponent, he loses immediately. If no one loses immediately (in this way), they ultimately play an ω-sequence of ordinals, and the payoff is that I wins this play of the game iff the sup of the sequence of ordinals just played is a member of Q. This is certainly not the sort of game considered when we defined AD, but

for the sake of motivation, let us assume that one of the players, say player I, has a winning strategy for the game. Then we claim that Q contains an ω-closed unbounded subset (if we were to assume that II had the winning strategy, then Q^c would contain the ω-closed unbounded set). Here is the argument: let f, a map from finite sequences of ordinals less than $\aleph_1$ into $\aleph_1$, be I's strategy for the game. Since we are assuming countable choice, $\aleph_1$ is a regular cardinal, and so given any α less than $\aleph_1$, the range of f on all finite sequences of ordinals less than α is bounded below $\aleph_1$. Given any α, let b_α be the least such bound.

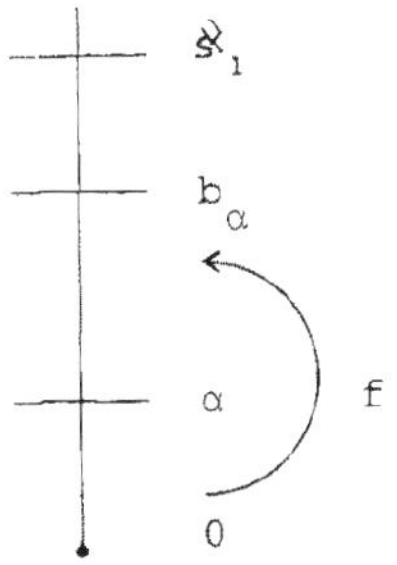

Let, now, γ be the sup of the ordinals $0, b_0, b_{b_0}, \ldots$. Then it is clear that the range of f on finite sequences of ordinals less than γ is bounded by γ. Let us call such ordinals, ordinals α such that $b_\alpha \leq \alpha$, closure points of f. It is clear from our construction of γ that the set of closure points for f, P, is unbounded in $\aleph_1$, and so it is immediate that the collection of sups of ω-sequences from P, $(P)_\omega$, is an ω-closed unbounded subset of $\aleph_1$.

Claim: $(P)_\omega \subseteq Q$. (Proof of claim: Suppose $\alpha_1, \alpha_2, \alpha_3, \ldots$ is an ω-sequence from P. Then since each α_i is a closure point of f, the play of our game which results when player I uses the strategy f and player II, for his i^{th} move always writes α_i, has sup $\bigcup_{i<\omega} \alpha_i$. Thus as f is a winning strategy for player I, we must have $\bigcup_{i<\omega} \alpha_i \in Q$. ⊠). □

This argument just completed begins to show how games can be used in producing closed unbounded sets. However, in order to use AD, we cannot rely on ordinal games. We must use integer games, and so a further idea is needed:

<u>Theorem 2.2</u> (Solovay): Assuming AD, the filter of sets containing ω-closed unbounded subsets of $\aleph_1$ is an ultrafilter.

<u>Proof</u>: We shall follow the above sketch (that in terms of ordinal games) quite closely. First off, let us fix once and for all some standard coding of countable ordinals via reals (members of ${}^{\omega}\omega$). There are many ways to do this since a countable well-ordering can be viewed simply as a countable set of ordered pairs of nonnegative integers, the set of ordered pairs just happening to entail a well-ordering, and any standard recursive pairing function can now be used to code a set of ordered pairs of natural numbers into a single member of ${}^{\omega}\omega$.

At any rate, given a single member f of ${}^{\omega}\omega$, let, for each $n < \omega$, f_n denote the member of ${}^{\omega}\omega$ given by $f_n(m) = f(2^n 3^m)$ for all m. Then in complete analogy to our ordinal game sketch given previously we proceed as follows:

> suppose A is a given subset of $\aleph_1$. We wish to prove that either A or A^c contains an ω-closed unbounded subset of $\aleph_1$, and so let us consider the following <u>integer</u> game played between our two players I and II: I and II move alternately starting with I, writing at each move a

member of ω. After ω many moves, I and II have each produced a member of ${}^{\omega}\omega$. Say I produces f and II produces g. Then the payoff is this: If for some n, f_n or g_n fails to code a countable ordinal under our standard coding, then the player who first failed loses (i.e., let n be the least n such that f_n or g_n does not code a well-ordering. If f_n does not code a well ordering, I loses – otherwise II loses). Assume, on the other hand, that for each n, f_n and g_n code well orderings α_n and β_n, and let α be the sup of all such ordinals α_n and β_n. Then I wins the play iff $\alpha \in A$. Now although we have given a somewhat lengthy verbal description of this game, it is easily seen to be a standrad integer game, and so by AD, let s be a winning strategy for one of the players.

Assume that s is a winning strategy for player I. We will show that that A contains an ω-closed unbounded subset. (If s were a winning strategy for player II, a parallel argument would show A^c to contain the ω-closed unbounded subset). Now in analogy with our previous argument using ordinal games, we wish to have a notion of "bounding ordinal", i.e., a process for going from α to b_α as we had earlier. So let, for any $\alpha < \aleph_1$, X_α be the set of reals which result when the strategy s is played against reals coding finite sequences sitting below α, that is, if $s*g$ denotes the result of a play of s against g, then

$$X_\alpha = \{(s*g)_n \mid g_m \text{ codes an ordinal less than } \alpha \text{ for all } m < n\}.$$

It is easy to see that X_α is a Σ^1_1 set of reals coding ordinals, and so by the Kleene hierarchy theorem, there exists <u>a countable</u> ordinal β_α such that $r \in X_\alpha$ implies that the ordinal coded by r is less than β_α.

We can now complete our argument as we did in the above sketch which used ordinal games. Our bounding ordinals β_α work in this context exactly as

the b_α worked previously, and
the proof proceeds exactly as before.

□

<u>Corollary 2.3</u> (Solovay): Assuming AD, $\aleph_1$ is a measurable cardinal.

<u>Proof</u>: The result here follows immediately from 2.2 and 1.1. For since we are working in AD + DC, 1.1 tells us that the measure on $\aleph_1$ associated with the ultrafilter $W_{\aleph_1}$ is $\aleph_1$-additive (i.e., the union of any fewer-than-$\aleph_1$-many sets of measure 0 has measure 0). Thus $\aleph_1$ is measurable. □

This last result sparked modern interest in AD. It instigated a great deal of study of the axiom, but nothing substantially new was forthcoming until Martin broke the ice with an ingenious proof. It begins with a lemma:

<u>Lemma 2.4</u> (Martin): Assume AD. Then any set of Turing degrees either contains a cone or is disjoint from a cone (A cone is a set of degrees of the form $\{\underset{\sim}{d} \mid \underset{\sim}{d} \geq \underset{\sim}{d}_0\}$).

<u>Proof</u>: Suppose $\mathcal{E}$ is a given set of (Turing) degrees. Consider the set of reals E whose degree is in $\mathcal{E}$. By AD,

the game G_E is determinate, and so let f be an associated winning strategy. Then it is immediate that if $\underset{\sim}{d}_0$ is the Turing degree of f (keep in mind: f is just a function from finite sequences of integers to integers), then $\underset{\sim}{d}_0$ is a vertex of our desired cone. Indeed, if f is a winning strategy for player I, then $\{\underset{\sim}{d} \mid \underset{\sim}{d} \geq \underset{\sim}{d}_0\} \subseteq E$, for given $\underset{\sim}{d} \geq \underset{\sim}{d}_0$, if II plays a real of degree $\underset{\sim}{d}$ and I plays using f, the result is a real of degree $\underset{\sim}{d}$ which is a win for I in G_E. Thus $\underset{\sim}{d} \in E$. Similarly, if f is a winning strategy for player II, then $\{\underset{\sim}{d} \mid \underset{\sim}{d} \geq \underset{\sim}{d}_0\} \subseteq E^c$.

□

<u>Theorem 2.5</u>: Assuming AD, $\aleph_1$ is a measurable cardinal.

<u>Proof</u> (Martin): Let $\mathfrak{S}$ be the set of all Turing degrees. Then we can define a nontrivial two-valued measure μ on $\mathfrak{S}$ by

"$\mu(E) = 1$ iff E contains a cone."

Since any countable set of Turing degrees has a sup, μ is clearly countably additive. We can now translate μ into a measure ν on $\aleph_1$ as follows:

> let g be the function which sends any degree to the least ordinal not recursive in it. Then we can define ν by

$$\nu(A) = 1 \quad \text{iff} \quad \mu(g^{-1}(A)) = 1.$$

Since μ is countably additive, so is ν. It is immediate that μ is nontrivial, and so $\aleph_1$ must be measurable. □

Clearly Martin's proof from AD of the measurability of $\aleph_1$ is far simpler than Solovay's. At first glance, it may seem that Solovay's argument yields a stronger result, for in our proof of 2.5 the measure constructed is not the one generated by the filter of ω-closed unbounded sets. However, with a slight modification of our argument in 2.5 (simply use "stable" ordinals in place of "admissible" ordinals), the stronger result follows equally easily.

Martin's ideas here turned out to be very powerful. Among other things, they led to Solovay's original proof from AD that $\aleph_2$ is a measurable cardinal. We do not wish to concentrate on this result here (we will give an elegant proof of it in Chapter IV), but rather wish to examine another consequence of Martin's short proof of 2.5, the existence, from AD, of infinite exponent partition relations:

Theorem 2.6 (Martin): Assuming AD, $\aleph_1 \longrightarrow (\aleph_1)^{\omega}$.

Proof: We must begin with a lemma of Jensen which strengthens Sacks' celebrated coding theorem for admissible ordinals:

<u>Lemma</u> (Jensen): Assume $\alpha_1, \alpha_2, \ldots.$ is an ω-sequence of ordinals each of which is admissible in a fixed degree $\underset{\sim}{d}_0$. Then there is a degree $\underset{\sim}{d} \geq \underset{\sim}{d}_0$ such that $\alpha_1, \alpha_2, \ldots.$ are the first ω-many admissibles in $\underset{\sim}{d}$.

Jensen's lemma can be proved using either almost-disjoint set forcing or an appropriate Barwise compactness argument. We will not prove this lemma here, but rather will proceed directly with the proof of 2.6: given a partition $F: [\aleph_1]^{\omega} \longrightarrow 2$, let E be the set of degrees $\underset{\sim}{d}$ such that if $\alpha_1^{\underset{\sim}{d}}, \alpha_2^{\underset{\sim}{d}}, \ldots.$ is the sequence of the first ω-many ordinals admissible in $\underset{\sim}{d}$, then $F(\{\alpha_1^{\underset{\sim}{d}}, \alpha_2^{\underset{\sim}{d}}, \ldots.\}) = 0$. Then by lemma 2.4, either E or E^c contains a cone. Suppose $\{\underset{\sim}{d} \mid \underset{\sim}{d} \geq \underset{\sim}{d}_0\} \subseteq E$ (a similar argument holds if E^c contains the cone). Let C be the set of all ordinals admissible in $\underset{\sim}{d}_0$. Then we claim that C is homogeneous for F, and in fact that $F''[C]^{\omega} = \{0\}$. For if $\alpha_1, \alpha_2, \ldots.$ is a member of $[C]^{\omega}$, Jensen's lemma says there is a $\underset{\sim}{d} \geq \underset{\sim}{d}_0$ such that $\alpha_1, \alpha_2, \ldots.$ are the first ω-many $\underset{\sim}{d}$-admissibles. But as $\{\underset{\sim}{d} \mid \underset{\sim}{d} \geq \underset{\sim}{d}_0\} \subseteq E$, the definition of E tells us that $F(\{\alpha_1, \alpha_2, \ldots.\}) = 0$. □

Following Martin's discovery of 2.6, a great deal of effort was expended attempting to prove partition relations for $\aleph_1$ with exponent greater than ω. Finally, Kunen was able to show AD to imply $\aleph_1 \longrightarrow (\aleph_1)^{\alpha}$ for all countable α. His proof, however, used iterated ultraproducts, was extremely delicate, and left little hope for further progress.

Once again, a simple observation by Martin broke the way for future work. Martin belatedly realized that Solovay's original proof in 2.2 could be modified slightly to yield Kunen's theorem. Keeping in mind our motivational sketch prior to 2.2, we might describe Martin's idea as follows: suppose we are given an $\alpha < \aleph_1$ and a partition $F : [\aleph_1]^{\alpha} \longrightarrow 2$. Then instead of thinking of the two players I and II as building a single ω-sequence of ordinals during a play of the appropriate associated game, think of them as building infinitely many such sequences. When they are done, instead of looking at the sup of the single sequence, look at the infinitely many sups of all the sequences constructed. Let us call $\alpha_1, \alpha_2, \ldots, \alpha_\beta, \ldots$ these sups, and if $\{\alpha_1, \alpha_2, \ldots, \alpha_\beta, \ldots\}$ has order-type α (we can arrange the game to require this), then we simply say that player I wins this play iff $F(\{\alpha_1, \alpha_2, \ldots, \alpha_\beta, \ldots\}) = 0$. Other than this one change, that of playing infinitely many sequences (all of which can clearly be coded into a single play of a G_A-type game), there is little difference between the proof here and that of 2.2. Given the set Q of closure points of a winning strategy for such a game, we get a homogeneous set for F simply by taking it to be the set of sups of all those sequences from Q which consist of ω-many members of Q in a row (i.e., if q_α denotes the αth member of Q, our desired homogeneous set would be $\left\{\bigcup_{n<\omega} q_{\omega\cdot\alpha+n} \middle| \alpha < \aleph_1\right\}$).

(Note: In the proof of 2.2, we took our closed unbounded subset of the given set to be the set of sups of <u>all</u> ω-sequences from Q —— if you check the details in the case of partitions with exponent greater than ω, the set of sups of <u>all</u> ω-sequences from Q is not generally homogeneous.)

We will not give a detailed rendition of Martin's proof of Kunen's theorem here because Martin subsequently proved an even better theorem. It is this theorem that is the main goal of this chapter.

Theorem 2.7 (Martin): Assuming AD, $\aleph_1 \longrightarrow (\aleph_1)^{\aleph_1}$.

The proof of 2.7 follows, in a sense, the sketch of Martin's proof of Kunen's theorem given just above. The only difficulty is that although a game with countably many integer moves can code arbitrary countable sequences of countable ordinals, how can such a game code an arbitrary uncountable sequence? The answer comes via an early result of Solovay.

Lemma 2.8 (Solovay): Assuming AD, every subset of $\aleph_1$ is constructible from a real. In fact, there is a single formula $\Phi(x,y)$ of set theory with the following property: for any subset A of $\aleph_1$, there is a real r such that for any $\alpha < \aleph_1$,

$$\alpha \in A \quad \text{iff} \quad L[r] \models \Phi(\alpha,r).$$

Proof: Suppose A is a given subset of $\aleph_1$. Then we can consider a game played between two players, I and II, as follows: I and II move alternately writing nonnegative integers. Let p be the member of ${}^{\omega}\omega$ so built by I, and q that built by II. Then II wins this play of the game iff p fails to code a countable well-ordering (see proof of 2.2), or p does code a well-ordering, say α_0, but q codes the countable set of ordinals $A \cap \alpha$ for some $\alpha \geq \alpha_0$.

Now by using an idea from the proof of 2.2, it is fairly easy to see that player I can never have a winning strategy for this game. For if s is a winning strategy for player I, then that set S of all members of ${}^{\omega}\omega$ built by using the strategy s during all possible plays of our game is a Σ^1_1 set of reals coding ordinals. Thus, by the Kleene hierarchy theorem, there exists a countable ordinal α_S such that $r \in S$ implies that the ordinal coded by r is less than α_S. Our contradiction is now quite close. For if q is any real which codes the countable set of ordinals $A \cap \alpha_S$, player II playing q as his moves against I's playing via the strategy s is a win for II, thereby contradicting the supposition that s is a winning strategy for I.

Given now that player I can never have a winning strategy for the game just described, AD tells us that player II must have a winning strategy. Let t be such a strategy. Then we claim that A is constructible from t. This would be immediate if $\aleph_1$ and $\aleph_1^{L[t]}$ were the same, for in that case, we would be able (within $L[t]$) to see if a countable ordinal α were in A by simply playing t against I's playing a code for α and then decoding the result. However, if $\aleph_1^{L[t]} < \aleph_1$, there will be countable ordinals not coded by reals in L[t]. In this case, we must use generic codings, and so we must appeal to a forcing argument: it is clear from discussion above that there is a Σ^1_2 formula $\theta(r,s)$ such that for any real p, p codes an ordinal A if $\theta(p,t)$. For any ordinal α, let P^α be the standard partial ordering which generically counts α:

$$P^\alpha =_{df} \langle \{f \mid \text{ for some } n < \omega, \ f:n \longrightarrow \alpha\}, \subseteq \rangle .$$

Since $\aleph_1$ is measurable, it is strongly inaccessible in L[t]. Thus, $2^{P^\alpha} \cap L[t]$ is countable for every countable α. This tells us that for any $\alpha < \aleph_1$, there always exists an L[t] - generic filter on P^α. Thus, for any $\alpha < \aleph_1$, if we let $\underline{\tau_\alpha}$ denote a term for a real coding α in every P^α-generic extension of L[t], we have

$$\alpha \in A \quad \text{iff} \quad \text{for every } L[t]\text{-generic } G \text{ over } P^\alpha, \ L[t,G] \models \theta(\tau_\alpha, t)$$

$$\text{iff} \quad L[t] \models \phi \Vdash \theta(\underline{\tau_\alpha}, \underline{t}) .$$

□

<u>Proof of 2.7</u>: Let us first introduce some notation: if Q is any unbounded subset of $\aleph_1$, let ${}_\omega Q$ denote the collection of sups of every sequence from Q which consists of ω-many members of Q in a row, that is, if q_α denotes the αth largest member of Q then

$${}_\omega Q =_{df} \left\{ \bigcup_{n<\omega} q_{\omega\cdot\alpha+n} \mid \alpha < \aleph_1 \right\} .$$

In our proof we shall make strong use of Silver's theory of indiscernibles for L, and so let us introduce some associated notation: in the usual rendition, the Π^1_2 sentence "$y = x^\#$" is a conjuction of two parts, a Δ^1_1 sentence $I(x,y)$ stating that y codes a theory extending $ZFC + V = L[x]$ having ω-many indiscernibles and containing certain special "combinatorial" sentences, and a Π^1_2 sentence

$$\text{"}\forall\alpha < \aleph_1 \left(M(\alpha,x,y) \text{ is well-founded"}\right),$$

where $M(\alpha,x,y)$ denotes the minimal model for the theory y generated by α-many indiscernibles. Let $WF(\alpha,x,y)$ denote the order-type of the well-founded portion of $M(\alpha,x,y)$, that is, the order-type of the set of well-founded ordinals of $M(\alpha,x,y)$.

We are now ready to proceed with the proof: the fundamental idea is to follow our sketch just given for proving $\aleph_1 \longrightarrow (\aleph_1)^\alpha$, but here, instead of using a simple method for coding arbitrary α-sequences of countable ordinals within countably many moves of a game, using Lemma 2.8 to code arbitrary <u>$\aleph_1$-sequences</u> of countable ordinals with only countably many moves. In order that this coding work, we will have to have strong control of L, and this is where Silver's theory of indiscernibles comes in. Here are the details: suppose we are given a partition $F:[\aleph_1]^{\aleph_1} \longrightarrow 2$. Then we will find a closed unbounded subset Q of $\aleph_1$ such that ${}_\omega Q$ is homogeneous for F.

Let $\{\Phi_i\}_{i<\omega}$ be an enumeration of all formulas of the language of set theory, and let, for each i, the function $f_i^x(y)$ be given by

$$f_i^x(z) = \begin{cases} \text{the least ordinal } \alpha \text{ such that } \Phi_i(x,\alpha,z) \text{ if } \Phi_i \text{ is a formula of three variables and such an } \alpha \text{ exists} \\ 0 \quad \text{otherwise.} \end{cases}$$

Let $S(\alpha,n,x,y)$ denote

"$I(x,y)$ and $\alpha < WF(\alpha+1,x,y)$ and $\left(f_n^x{}''[\alpha+1]\right)^{M(\alpha+1,x,y)} \subseteq WF(\alpha+1,x,y)$".

Loosely speaking, $S(\alpha,n,x,y)$, says that although $M(\alpha+1,x,y)$ might not be well-founded, the function f_n^x, when viewed within $M(\alpha+1,x,y)$, sends all ordinals $\leq \alpha$ to within the well-founded portion. We can now consider a game played between two players, I and II, as follows: using a standard fixed coding, let us decode any play r of the game into an integer n_1 and two reals x_1 and y_1 played by I and an integer n_2 and two reals x_2 and y_2 played by II. Then we say I wins this play of the game iff either:

a) $S(\beta,n_1,x_1,y_1)$ yet $\neg S(\alpha,n_2,x_2,y_2)$ for some $\alpha < \beta < \aleph_1$,

or

b) $y_1 = x_1^{\#}$ and $y_2 = x_2^{\#}$ and the following function of α, a member of $[\aleph_1]^{\aleph_1}$, is sent by the partition F to 0:

$$\sup_{i<\omega} \left\{ f_{n_1}^{x_1}(\omega\cdot\alpha+i)^{L[x_1]},\ f_{n_2}^{x_2}(\omega\cdot\alpha+i)^{L[x_2]} \right\}.$$

(Intuitively, I wins a play of this game iff he kept more of the range of $f_{n_1}^{x_1}$ in a well-founded portion of his approximation to $L[x_1]$ than II did with x_2 for his approximation to $L[x_2]$

<u>or</u>

the play suitably coded a member of $[\aleph_1]^{\aleph_1}$ sent by F to 0.)

By AD, there exists a winning strategy for this game. Since the game is clearly symmetric, we can assume without loss of generality that it is I who has the winning strategy. Let s denote such a strategy, and let us denote by $s(n)$, $s(x)$, $s(y)$ the decoded version of I's moves when he uses s against II playing n, x, y.

As in the proof of 2.2, we wish to find "bounding ordinals" for our strategy s, and although the details are slightly messier here, the idea is basically the same: let, for α and β ordinals and $i < \omega$, $X_{\alpha,i}^{\beta}$ denote

$\{ r \mid r$ codes a countable ordinal τ with the following property: for some n, x, and y, $I(x,y)$, and for every $\delta < \omega \cdot \alpha + i$,

1) $M(\delta+1,x,y)$ is well-founded or $\beta \leq WF(\delta+1,x,y)$,

2) $f_n^x(\delta)^{M(\delta+1,x,y)}$ is a standard ordinal $< \beta$, and

3) $\tau \leq f_{\mathfrak{s}(n)}^{\mathfrak{s}(x)}(\omega \cdot \alpha + i)^{M(\omega\cdot\alpha+i+1,\mathfrak{s}(x),\mathfrak{s}(y))} \}$.

For each α, β and i, $X_{\alpha,i}^{\beta}$ is a Σ_1^1 set of reals coding ordinals, and so by the Kleene hierarchy theorem let, for each α, β and i, $\sigma_{\alpha,i}^{\beta}$ be the least <u>countable</u> ordinal such that

$$r \in X_{\alpha,i}^{\beta} \implies r \text{ codes an ordinal } < \sigma_{\alpha,i}^{\beta}.$$

Our proof is now almost complete. Let Q denote

$$\{ \eta < \aleph_1 \mid \text{for every } \alpha, \beta \text{ and } i, \; \omega \cdot \alpha + i \leq \beta < \eta \text{ implies } \sigma_{\alpha,i}^{\beta} < \eta \}.$$

Q is our set of closure ordinals, and we claim that ${}_{\omega}Q$ is homogeneous for F. In fact,

Claim: $F''[{}_{\omega}Q]^{\aleph_1} = \{0\}$.

(Proof of claim: Suppose $p \in [{}_{\omega}Q]^{\aleph_1}$. Then $p = {}_{\omega}g$ for some $g \in [Q]^{\aleph_1}$. By our coding lemma 2.8, let $n < \omega$ and $x \leq \omega$ be such that $g(\alpha) = f_n^x(\alpha)^{L[x]}$ for each $\alpha < \aleph_1$. Then we consider a play of our game in which II plays n, x and $x^{\#}$, and I plays $\mathfrak{s}$. Now since $\mathfrak{s}$ is a winning strategy for I, and since II constructed $x^{\#}$ correctly during his moves, we must have

1) $s(x^{\#}) = s(x)^{\#}$, and

2) if h is the member of $[\aleph_1]^{\aleph_1}$ given by

$$h(\alpha) = \sup_{i<\omega} \left\{ f^{s(x)}_{s(n)}(\omega\cdot\alpha+i)^{L[s(x)]},\ f^{x}_{n}(\omega\cdot\alpha+i)^{L[x]} \right\},$$

then $F(h) = 0$.

Our result will thus be proved if we could show that $h = {}_{\omega}g$. This we now do: we first note that

$f^{s(x)}_{s(n)}(\omega\cdot\alpha+i)^{L[s(x)]} < \sigma^{g(\omega\cdot\alpha+i)}_{\alpha,i}$ for every $\alpha < \aleph_1$ and $i < \omega$. This follows simply by checking the definition of $x^{g(\omega\cdot\alpha+i)}_{\alpha,i}$ using $n, x, x^{\#}$ for n, x, y —— $I(x, x^{\#})$ is true, for every $\delta < \omega\cdot\alpha + i$ $M(\delta+1,x,x^{\#})$ is well-founded, and since it is an elementary substructure of $L[x]$, $f^{x}_{n}(\delta)^{M(\delta+1,x,x^{\#})} = f^{x}_{n}(\delta)^{L[x]} = g(\delta) < g(\omega\cdot\alpha+i)$, and finally, $f^{s(x)}_{s(n)}(\omega\cdot\alpha+i)^{M(\omega\cdot\alpha+i+1,s(x),s(x^{\#}))}$ is equal to $f^{s(x)}_{s(n)}(\omega\cdot\alpha+i)^{M(\omega\cdot\alpha+i+1,s(x),s(x)^{\#})}$ (as $s(x^{\#}) = s(x)^{\#}$), and so $f^{s(x)}_{s(n)}(\omega\cdot\alpha+i)^{M(\omega\cdot\alpha+i+1,s(x)s(x^{\#}))}$ equals $f^{s(x)}_{s(n)}(\omega\cdot\alpha+i)^{L[s(x)]}$. Next note that for any $\alpha < \aleph_1$, since $\omega\cdot\alpha+i \leq g(\omega\cdot\alpha+i) < {}_{\omega}g(\alpha)$, the fact that ${}_{\omega}g(\alpha) \in Q$ tells us $\sigma^{g(\omega\cdot\alpha+i)}_{\alpha,i} < {}_{\omega}g(\alpha)$. Thus for every $\alpha < \aleph_1$, $f^{s(x)}_{s(n)}(\omega\cdot\alpha+i)^{L[s(x)]} < {}_{\omega}g(\alpha)$. From the definition of h and from the fact that $g(\alpha)$ always equals $f_n^x(\alpha)^{L[x]}$, this latest fact gives that $h =_\omega g$, and so our theorem is proved.)

□

We need one further consequence of AD before we can proceed:

Consider the ultrapower

$$\aleph_1^{\aleph_1}/\mu$$

where μ is the measure on $\aleph_1$ generated by the filter of ω-closed unbounded sets (we are assuming AD, and so such a measure exists by 2.2). Since we have dependent choice at our disposal, we know that this ultrapower is well-ordered, and so the question arises as to its order-type. A result of Solovay answers the question:

<u>Theorem 2.9</u> (Solovay): Assuming AD, $\aleph_1^{\aleph_1}/\mu$ has order-type $\aleph_2$ (where μ is the measure on $\aleph_1$ generated by the filter of ω-closed unbound sets.)

<u>Proof</u>: There are a number of different ways to see that $\aleph_1^{\aleph_1}/\mu \geq \aleph_2$ and, in fact, that if κ is any measurable cardinal with measure μ, then

$$\kappa^\kappa/\mu \geq \kappa^+.$$

The simplest way is to explicitly construct, for each $\beta < \kappa^+$, a β-sequence of functions f_α in $[\kappa]^\kappa$ with the property that $\alpha_1 < \alpha_2 < \beta$ implies $f_{\alpha_1} < f_{\alpha_2}$ almost everywhere (μ). Since we do this in Chapter IV (specifically in Lemma 4.2), we will not do it again here.

The difficult thing to prove is that $\aleph_1^{\aleph_1}/\mu \leq \aleph_2$, and for this we are forced to rely upon Corollary 2.3 and Lemma 2.8. Indeed, let $\theta(x,y,z)$ be our formula from 2.8 such that for any map f from $\aleph_1$ into $\aleph_1$, there exists a real t such that

$$f(\alpha) = \beta \quad \text{iff} \quad L[t] \models \theta(\alpha,\beta,t)$$

for every α and β below $\aleph_1$. By Corollary 2.3, there exists a measurable cardinal, and hence, we may apply Silver's theory of indiscernibles.

Suppose that $f: \aleph_1 \longrightarrow \aleph_1$ is a given map and that t is a real such that

$$f(\alpha) = \beta \quad \text{iff} \quad L[t] \models \theta(\alpha,\beta,t)$$
$$\text{for every } \alpha \text{ and } \beta \text{ below } \aleph_1.$$

Then as $L[t] \models \forall\alpha < \aleph_1 \, \exists!\beta < \aleph_1 \theta(\alpha,\beta,t)$ and as $\aleph_1$ and $\aleph_2$ are indiscernibles,

$$L[t] \models \forall\alpha < \aleph_2 \, \exists!\beta < \aleph_2 \theta(\alpha,\beta,t).$$

Let us denote by β_f^t the unique β below $\aleph_2$ such that

$$L[t] \models \theta(\aleph_1, \beta, t).$$

Claim: For any f and g in $\aleph_1^{\aleph_1}$,
if $f(\alpha) = \beta$ iff $L[t_1] \models \theta(\alpha, \beta, t_1)$ for all $\alpha, \beta < \aleph_1$,
and $g(\alpha) = \beta$ iff $L[t_2] \models \theta(\alpha, \beta, t_2)$ for all $\alpha, \beta < \aleph_1$,
then $f < g$ a.e. (μ) iff $\beta_f^{t_1} < \beta_g^{t_2}$.

(Proof of claim: Work in $L[t_1$ join $t_2]$: it is easy to see that since $\aleph_1$ is an $L[t_1$ join $t_2]$-Silver-indiscernible, $\beta_f^{t_1} < \beta_g^{t_2}$ iff the unique β such that $L[t_1] \models \theta(\alpha, \beta, t_1)$ is less than the unique β such that $L[t_2] \models \theta(\alpha, \beta, t_2)$ for <u>every</u> $L[t_1$ join $t_2]$-Silver-indiscernible α. Since the set of countable $L[t_1$ join $t_2]$-Silver-indiscernibles is closed and unbounded below $\aleph_1$, it has μ-measure 1, and hence $\beta_f^{t_1} < \beta_g^{t_2}$ iff $f < g$ a.e. (μ). $\boxtimes$)

By this claim, we now immediately have that the map which sends any f in $\aleph_1^{\aleph_1}$ to β_f^t (where t is such that

$$f(\alpha) = \beta \text{ iff } L[t] \models \theta(\alpha, \beta, t) \text{ for all } \alpha, \beta < \aleph_1),$$

is independent of which t we choose and extends to a well-defined map of

$\aleph_1^{\aleph_1}/\mu$ order-preservingly into $\aleph_2$.

We thus have that

$\aleph_1^{\aleph_1}/\mu \leq \aleph_2$, and our result is proved. $\square$

In subsequent chapters, we shall take these last two consequences of AD, $\aleph_1 \longrightarrow (\aleph_1)^{\aleph_1}$ and $\aleph_1^{\aleph_1}/\mu \simeq \aleph_2$, and <u>derive</u> <u>from</u> <u>them</u> <u>alone</u> all combinatorial consequences of AD below $\aleph_\omega$.

Chapter III

Let us now reexamine our results from the second half of Chapter I, this time in more detail, and see to just what extent infinite exponent partition properties of γ imply measurability properties of γ. Throughout this chapter, we work only in ZF.

Our first goal is to derive the partition relations used in Chapter I to show γ measurable from something more natural. Here is the result.

Lemma 3.1: Assume $\gamma \longrightarrow (\gamma)^{\omega+\omega}$. Then for every $\lambda < \gamma$, $\gamma \longrightarrow (\gamma)^{\omega}_{\lambda}$.

Proof: Suppose $F : [\gamma]^{\omega} \longrightarrow \lambda$ is a given partition. Then we define an auxilliary partition $G : [\gamma]^{\omega+\omega} \to 2$ by

$$G(x) = 0 \quad \text{iff} \quad F(x_0) = F(x_1),$$

where x_0 denotes the first ω-many members of x and x_1 denotes the second ω-many. Let C be homogeneous for G in the sense that $G''[C]^{\omega+\omega} = 1$. Claim: $G''[C]^{\omega+\omega} = \{0\}$. (Proof: Let y_1 denote the first ω-many members of C, y_2 denote the second ω-many members, y_3 the third, and in general, y_α the αth. Then for each $\alpha < \kappa$, $y_\alpha \in [C]^{\omega}$, and if $\alpha < \beta < \kappa$, then $y_\alpha \cup y_\beta \in [C]^{\omega+\omega}$. As γ is a cardinal and $\lambda < \gamma$, the map g from γ into λ given by $g(\alpha) = F(y_\alpha)$ must fail to be 1-1, and so let α and β be such that $F(y_\alpha) = F(y_\beta)$. Then $y_\alpha \cup y_\beta \in [C]^{\omega+\omega}$ and $G(y_\alpha \cup y_\beta) = 0$. As C is homogeneous, $G''[C]^{\omega+\omega} = \{0\}$. $\square$).

It is now fairly easy to check that C is homogeneous for the original partition F. For given x_1 and x_2 in $[C]^\omega$, let y be a member of $[C]^\omega$ whose least element exceeds the sup of x_1 and x_2. (Such a y can be found as $\gamma \longrightarrow (\gamma)^{\omega+\omega}$ implies $\gamma \longrightarrow (\gamma)^2$ which in turn implies that γ is a regular cardinal.) As $x_1 \cup y$ and $x_2 \cup y$ are both in $[C]^{\omega+\omega}$, our previous claim tells us that $G(x_1 \cup y) = G(x_2 \cup y) = 0$. Hence by the definition of G, $F(x_1) = F(y) = F(x_2)$. $\square$

By this lemma, the partition relation on $\aleph_1$ given us by AD, $\aleph_1 \longrightarrow (\aleph_1)^{\aleph_1}$, is enough in itself to imply that $W_{\aleph_1}$ is a nontrivial countably additive normal measure on $\aleph_1$. ($\aleph_1 \longrightarrow (\aleph_1)^{\aleph_1}$ of course implies $\aleph_1 \longrightarrow (\aleph_1)^{\omega+\omega}$.)

—————— o ——————

Thus far we have restricted our attention to the notion of "ω-closed" sets, but this is not at all necessary. Given any limit ordinal η less than our regular uncountable cardinal γ, we can define a subset Q of γ to be η-closed if the sup of every η-sequence of members of Q is a member of Q. With virtually no change, our results of Chapter I carry over. Lemma 3.1 remains true as well, and we can summarize as follows:

Theorem 3.2: Assume η is a limit ordinal less than γ and that γ satisfies $\gamma \longrightarrow (\gamma)^{\eta+\eta}$. Then the measure on γ which gives a set measure 1 iff it contains an η-closed unbounded subset is a nontrivial two-valued γ-additive normal measure on γ.

Proof: Theorem 1.10 and Lemma 3.1 routinely generalize to an arbitrary limit ordinal η in place of ω. These generalized versions of 1.10 and 3.1 immediately yield 3.2. □

Notation: Let the measure on γ described in the statement of 3.2 be denoted μ_η.

Let us now assume that γ satisfies $\gamma \longrightarrow (\gamma)^{\gamma}$. Then by 3.2, we have a nontrivial two-valued γ-additive normal measure μ_η on γ for each limit ordinal η less than γ. Are these measures really different for different η? The following two theorems give the answer:

Theorem 3.3: If η_1 and η_2 are distinct regular cardinals less than γ, then μ_{η_1} and μ_{η_2} are different measures.

Proof: Let $A_1 = \{\alpha < \gamma | \ cf(\alpha) = \eta_1\}$, and $A_2 = \{\alpha < \gamma | \ cf(\alpha) = \eta_2\}$. Clearly $\mu_{\eta_1}(A_1) = 1$ and $\mu_{\eta_2}(A_2) = 1$. But $A_1 \cap A_2 = \phi$, and so $\mu_{\eta_1}(A_2) = 0$. Thus $\mu_{\eta_1} \neq \mu_{\eta_2}$.

□

Theorem 3.4: For any limit ordinal η less than γ, μ_η and $\mu_{cf(\eta)}$ are the same measure.

Proof: Since any set which is $cf(\eta)$-closed is clearly η-closed, any set containing a $cf(\eta)$-closed unbounded subset also contains an η-closed unbounded subset. This tells us that for any $A \subseteq \gamma$, $\mu_{cf(\eta)}(A) = 1$ implies $\mu_\eta(A) = 1$. Since $\mu_{cf(\eta)}$ and μ_η are measures, we thus also have that for any $A \subseteq \gamma$, $\mu_{cf(\eta)}(A) = 0$ implies $\mu_{cf(\eta)}(A^c) = 1$ implies $\mu_\eta(A^c) = 1$ implies $\mu_\eta(A) = 0$. Thus $\mu_{cf(\eta)} = \mu_\eta$.

□

It is interesting to note that these previous two theorems remain valid even if we do not assume that the various μ_η are measures. In other words, for every limit ordinal η less than γ, let E_η denote the filter (not necessarily assumed to be ultra) of subsets of γ which contain η-closed unbounded subsets. Then the proof of 3.3 immediately yields that if η_1 and η_2 are distinct regular cardinals less than γ, $E_{\eta_1} \neq E_{\eta_2}$, and the proof of 3.4 immediately yields that for any limit η less than γ, $E_{cf(\eta)} \subseteq E_\eta$. But can we prove that for any limit $\eta < \gamma$ $E_{cf(\eta)} = E_\eta$ without assuming $E_{cf(\eta)}$ to be an ultra filter? The answer is yes, and it requires a bit of proof.

Fact: For any limit ordinal $\eta < \gamma$, $E_{cf(\eta)} = E_\eta$.
Proof (due to L. Tharp): From previous work, we know that $E_{cf(\eta)} \subseteq E_\eta$. Suppose $A \in E_\eta$, and let B be an η-closed unbounded subset of A. We will show that B contains a $cf(\eta)$-closed unbounded subset, and hence that A contains a $cf(\eta)$-closed unbounded subset. As A is arbitrary, we would then have our desired $E_\eta \subseteq E_{cf(\eta)}$. So how do we find a $cf(\eta)$-closed unbounded subset of B? Quite easily!
Claim: $(B)_\eta$ is a $cf(\eta)$-closed unbounded subset of B.
(Proof: Let τ be the least order-type of any final segment of η, that is, of any set of the form $\eta - \alpha$ for $\alpha < \eta$. Clearly, $\tau \geq cf(\eta) \geq \omega$. Let x denote a fixed sequence of order-type $cf(\eta)$ unbounded in τ. Now given a $cf(\eta)$ sequence y in $(B)_\eta$, we wish to show $\bigcup y \in (B)_\eta$. For every $\alpha < cf(\eta)$, let $y(\alpha)$ $(x(\alpha))$ be the αth largest element of y (x), and let z_α be the initial segment of $B - y(\alpha)$ of order-type equal to that of $\{\beta < \tau \mid x(\alpha) \leq \beta < x(\alpha + 1)\}$. Then $z =_{df} \bigcup_{\alpha < cf(\eta)} z_\alpha$ is a τ-sequence of elements of B whose sup equals that of y. Thus if z_{-1} is any sequence of elements of B of order-type equal to the least α such that $\eta - \alpha$ is order-isomorphic to τ whose sup does not exceed the least element of y, then $z_{-1} \cup z$ is an η-sequence of elements of B whose sup equals that of y. Thus $\bigcup y \in (B)_\eta$. $\boxtimes$)

$\square$

At times, these μ_η are the <u>only</u> normal measures on γ. Here is the result:

<u>Theorem 3.5</u>: Assume that for every limit $\eta < \gamma$, μ_η is a nontrivial two-valued γ-additive measure on γ. Then if there exist fewer than γ-many regular cardinals less than γ, the μ_η are the <u>only</u> normal nontrivial two-valued γ-additive measures on γ.

<u>Proof</u>: Let μ be a normal nontrivial 2-valued γ-additive measure on γ. Since μ is γ-additive, and since $\{cf(\alpha) \mid \alpha < \gamma\}$ has cardinality less than γ, $\{\alpha \mid cf(\alpha) = \lambda_0\}$ has μ-measure 1 for some $\lambda_0 < \gamma$. It is now fairly easy to see that $\mu = \mu_{\lambda_0}$. To see this, it suffices to show that for any $B \subseteq \gamma$, if $\mu_{\lambda_0}(B) = 1$, then $\mu(B) = 1$. Suppose $B \subseteq \gamma$, $\mu_{\lambda_0}(B) = 1$ yet $\mu(B) \neq 1$. We will derive a contradiction. As $\mu(B) \neq 1$, if we let $Q = \{\alpha \mid cf(\alpha) = \lambda_0\} \cap B^C$, then $\mu(Q) = 1$. Let C be a λ_0-closed unbounded subset of B. Then as $\alpha \in Q$ implies $\alpha \notin B$ and $cf(\alpha) = \lambda_0$, $\alpha \in Q$ implies $\bigcup(C \cap \alpha) < \alpha$. Thus if $g : Q \longrightarrow \gamma$ is given by $g(\alpha) = \bigcup(C \cap \alpha)$, the normality of μ tells us that for some $\alpha_0 < \gamma$, $\mu(\{\alpha \mid g(\alpha) = \alpha_0\}) = 1$. This clearly contradicts the fact that C is unbounded in γ.

□

Chapter IV

In this chapter we begin our work on cardinals beyond $\aleph_1$. Under the assumption of AD (we will really only use $\aleph_1 \longrightarrow (\aleph_1)^{\aleph_1}$ and $\aleph_1^{\aleph_1}/\mu_0 = \aleph_2$), we shall prove that $\aleph_2$ is a measurable cardinal having precisely two normal measures and that $\aleph_3$ is a Jonsson cardinal cofinal with $\aleph_2$. Our proofs here will be purely combinatorial and will introduce many of the techniques to be used in the next chapter in showing each $\aleph_n$ $(n \geq 2)$ Jonsson with cofinality $\aleph_2$. Throughout Chapters IV, V, and VI, we work in ZF + DC.

The following somewhat surprising result embodies the main technical facts upon which much of our work hinges. Although the theorem itself is not needed for our future work, its proof follows trivially from three lemmas each of which is in itself essential for our later work. The three lemmas are 4.2, 4.5, and 5.9. The proof of 4.1 will be given immediately following the proof of lemma 5.16.

Theorem 4.1: Let κ be a measurable cardinal and μ a normal measure on κ. Let f be a given member of $[\kappa]^\kappa$. Then given any subset Q of the ultrapower $[f]^\kappa/\mu$ such that $\bigcup Q - \bigcap Q$ has size at most κ, there exists a subset g of f such that Q is an initial segment of $[g]^\kappa/\mu$.

Proof: To be given following the proof of lemma 5.16.

Note: Theorem 4.1 is false if we do not at least assume $\bar{\bar{Q}} \leq \kappa$. This follows from cardinality considerations. (Assuming the axiom of choice, 4.1 remains valid with "$\bar{\bar{Q}} \leq \kappa$" in place of "$\overline{\overline{\bigcup Q - \bigcap Q}} \leq \kappa$".

Definition (and Convention): Let κ be a measurable cardinal and μ a nontrivial κ-additive measure on κ. Then by $<_\mu$ we mean the following linear ordering of $[\kappa]^\kappa$:

$$f <_\mu g \quad \text{iff} \quad \mu(\{\alpha \mid f(\alpha) < g(\alpha)\}) = 1.$$

Once a measure μ has been specified, we will always have this ordering in mind when viewing $[\kappa]^\kappa$ as an ordered set, so that, for example, by $[[\kappa]^\kappa]^\alpha$ we will mean the set of order preserving maps from $\langle\alpha,\varepsilon\rangle$ into $\langle[\kappa]^\kappa,<_\mu\rangle$.

Lemma 4.2 (Shuffling Lemma): Let κ be a measurable cardinal and μ a normal measure on κ. Then for every $\alpha < \kappa^+$, there exist functions $bk_\alpha\colon [\kappa]^\kappa \longrightarrow [[\kappa]^\kappa]^\alpha$ and $sf_\alpha\colon [[\kappa]^\kappa]^\alpha \longrightarrow [\kappa]^\kappa$ which are inverse to one another in the following sense: for every f in $[\kappa]^\kappa$, $sf_\alpha(bk_\alpha(f)) = f$, and for every F in $[[\kappa]^\kappa]^\alpha$, if $G = bk_\alpha(sf_\alpha(F))$, then for each $\beta < \alpha$, $G(\beta) = F(\beta)$ almost everywhere with respect to μ.

Proof: We begin by discussing a fairly simple consequence of the normality of μ which will be used in our proof.

Fact: Given any function $f : \kappa \longrightarrow [\kappa]^2$ such that for almost every α, $\bigcup f(\alpha) < \alpha$, there exists a pair $\{\alpha_1, \alpha_2\}$ in $[\kappa]^2$ such that for almost every α, $f(\alpha) = \{\alpha_1, \alpha_2\}$.
(Proof of fact: Let f_1 and f_2 mapping κ into κ be functions which satisfy $f(\alpha) = \{f_1(\alpha), f_2(\alpha)\}$ for all $\alpha < \kappa$. Then by the normality of μ there exist ordinals α_1 and α_2 and sets of measure 1 A_1 and A_2 such that $f_1''A_1 = \{\alpha_1\}$ and $f_2''A_2 = \{\alpha_2\}$. $A_1 \cap A_2$ is now a set of measure 1 and for every α in $A_1 \cap A_2$, $f(\alpha) = \{\alpha_1, \alpha_2\}$.)

We are now ready to prove the shuffling lemma. Our argument will be divided into 2 cases, $\alpha \geq \kappa$, and $\alpha < \kappa$.

$\alpha \geq \kappa$: Let h be a 1-1 map of κ onto α. When we refer to the "h-ordering on κ" we will mean the linear ordering of κ determined by h: $\eta_1 <_h \eta_2$ iff $h(\eta_1) \in h(\eta_2)$. Suppose we are given $f \in [\kappa]^\kappa$. We must describe $bk_\alpha(f)$, and for this we proceed as follows: imagine a square array of size κ by κ. Given f we will proceed to fill the lattice points in this square using, in order, ordinals from the range of f in addition to some other ordinals to be used as filler. We will fill the array inductively a row at a time (rows of the array will equivalently be referred to as "levels") and when we are done the κ-many columns will essentially constitute $bk_\alpha(f)$ — indeed, if for $\eta < \kappa$ f_η denotes the set of ordinals placed during our construction into the ηth column of the array, then $\beta \longmapsto f_{h^{-1}(\beta)}$ will be the map $bk_\alpha(f)$. Here now is how we distribute ordinals into our κ by κ array for $bk_\alpha(f)$: the 0th row of the array, all κ entries in it, are 0's — this is just filler. Proceeding inductively, suppose we have filled all rows below the νth, and in so doing have used up some initial segment of f. Then we fill the νth row as follows: the first ν-many entries of the νth row are filled using ordinals in the range of f which we haven't yet used. If τ_0 is the h-least ordinal among $\{0, 1, 2, 3, 4, 5, \ldots, \beta, \ldots\}_{\beta<\nu}$, then the least member of the range of f not yet used is placed in column τ_0 of row ν. If τ_1 is the second least (under $<_h$) ordinal less than ν, the next least member of the range of f not yet used is placed in column τ_1 of row ν. We continue in this way until all columns less than ν have received an entry on row ν. The remainder of row ν, those entries from the νth on, we simply fill with the ordinal ν. This is just filler. Proceeding in this way, we eventually build our array of ordinals, and it is routine to verify that if the map $f_\xi \in [\kappa]^\kappa$ is given by $f_\xi(\beta) =_{df}$ "the $<\xi,\beta>^{th}$ entry in array", then $\eta \longmapsto f_{h^{-1}(\eta)}$ is a member of $[[\kappa]^\kappa]^\alpha$. This member of $[[\kappa]^\kappa]^\alpha$ is $bk_\alpha(f)$, and since f was arbitrary, our description of bk_α is complete.

Now to describe sf_α: sf_α is basically the reverse of bk_α. Given $H \in [[\kappa]^\kappa]^\alpha$ we can arrange H so that its α-many κ-sequences are the columns in a κ by κ array. We can do this easily by using our map $h : \kappa \tilde{\to} \alpha$, for associated with H we shall consider the κ by κ array where the $<\delta,\eta>^{th}$ entry is $H(h(\delta))(\eta)$. We are going to build a member of $[\kappa]^{\kappa}$ by listing its elements in order, and we are going to choose the elements from among the entries in our κ by κ array determined by H. The member of $[\kappa]^{\kappa}$ we so build will be $sf_{\alpha}(H)$. Our construction of $sf_{\alpha}(H)$ will take place inductively by stages —— at the βth stage for $\beta < \kappa$, we will say how to add a β^*-sequence (some $\beta^* < \kappa$) of ordinals to the amount of $sf_{\alpha}(H)$ constructed to that point. This β^*-sequence of ordinals itself will be described inductively. Here is the construction: the 0th stage of our construction involves nothing. Now suppose we have completed the ηth stage of our construction for every $\eta < \beta$, and we wish to describe the βth stage. For this β^{th} stage, we will inductively pick a β^*-sequence (some $\beta^* < \kappa$) of ordinals from our κ by κ array determined by H each of which is larger than any ordinal we have put into $sf_{\alpha}(H)$ so far. Here is the first ordinal in the β^*-sequence: look at the first β columns of our array, and consider their ordering in the $<_{\mu}$-ordering. (β^* is the order-type of this ordering.) The ordinal we want is the least ordinal larger than everything yet used in building $sf_{\alpha}(H)$ which lies above row 0 and in the $<_{\mu}$-least column among the first β. Continuing, inductively on $\eta < \beta^*$, the ηth ordinal we add to $sf_{\alpha}(H)$ at stage β is the least ordinal larger than everything yet used which lies in the ηth $<_{\mu}$-least column among the first β columns. This describes the inductions, and $sf_{\alpha}(H)$ is the κ sequence from κ we so build. As H was arbitrary, we have defined sf_{α}.

Now it is routine to verify that for any $f \in [\kappa]^{\kappa}$,

$sf_\alpha(bk_\alpha(f)) = f$. What is harder is to show that for any H in $[[\kappa]^\kappa]^\alpha$, if $H' = bk_\alpha(sf_\alpha(H))$, then for all $\beta < \alpha$ $H'(\beta) = H(\beta)$ a.e. (μ). We do this now: let $g_1 : \kappa \longrightarrow \kappa$ as follows: $g_1(\beta) =_{df}$ "the least η such that during the ηth stage of the definition of $sf_\alpha(H)$, ordinals appearing in row β or higher of our κ by κ array associated with H are inserted into $sf_\alpha(H)$". It is easy to see that $g_1(\beta) \leq \beta$ for each β, and if $g_1(\beta) < \beta$ a.e. (μ), then by normality, this would mean that for some $\beta_o < \kappa$, $g_1(\beta) = \beta_o$ a.e. (μ). It is clear from our construction that this cannot be, and so we must have an A_1 such that $\mu(A_1) = 1$, and on A_1, $g_1(\beta) = \beta$. Next consider $g_2 : \kappa \longrightarrow [\kappa]^2$ given as follows: $g_2(\beta) =_{df}$ "either the least pair $\{\beta_1,\beta_2\}_<$ such that on row β of the κ by κ array for H the ordinal in column β_1 and the ordinal in column β_2 violate the h ordering, or $\{0,\beta\}$, whichever pair has smaller sup." Now if $\bigcup g_2(\beta) < \beta$ a.e. (μ), then by the fact given earlier in this proof, there would be a pair $\{\beta_1^o,\beta_2^o\}_<$ in $[\kappa]^2$ such that $g_2(\beta) = \{\beta_1^o,\beta_2^o\}$ a.e. (μ). But this would mean that the β_1^{oth} and β_2^{oth} columns of our κ by κ array for H violate the h ordering on full columns (i.e. $f <_h g$ iff $\mu(\{\alpha | f(\alpha) <_h g(\alpha)\}) = 1$), and this contradicts our very construction of the array. So let A_2 be such that $\mu(A_2) = 1$, and on A_2, $g_2(\beta) = \{0,\beta\}$, i.e., for any β in A_2, the first β many ordinals in the βth row preserve the h ordering on full columns.

It is clear from our construction that for any $\beta < \kappa$, during stage β of our definition of sf_α, the ordinals from our κ by κ array for H which are placed into our κ-sequence $sf_\alpha(H)$ are precisely the ordinals which would be placed into the first β-many columns of row β of the array for $bk_\alpha(sf_\alpha(H))$. Thus for any β in $A_1 \cap A_2$, the κ by κ arrays associated with H and with $bk_\alpha(sf_\alpha(H))$ have the same β^{th} row up through column β. Since $\mu(A_1 \cap A_2) = 1$, we must conclude that for every $\eta < \alpha$, $H(\eta)$ and $\big(bk_\alpha(sf_\alpha(H))\big)(\eta)$ are equal almost everywhere.

What about bk_α and sf_α in the cases where $\alpha < \kappa$? Here we proceed exactly as before except for the fact that our arrays will now have κ-many rows yet only α-many columns. Since $\alpha < \kappa$, we never have to worry about working above the diagonal — other than this, our construction and proof work as before.

□

Remark: The shuffling lemma just proved deals with functions which "break" and "shuffle" κ sequences of ordinals less than κ. By the same proof, however, the lemma remains valid in the more general context of breaking and shuffling κ-sequences from any well-ordered set of length at least κ. The only clause which must be added in the generalized context is one requiring that sequences being shuffled all have the same sup.

Before going on to give 4.5 (and hence to complete the proof of 4.1), we are in a position to prove the following result:

<u>Theorem 4.3</u>: Assume $\aleph_1 \longrightarrow (\aleph_1)^{\aleph_1}$ and $\aleph_1{}^{\aleph_1}/\mu_\omega = \aleph_2$. Then $\aleph_2 \longrightarrow (\aleph_2)^\alpha$ for each $\alpha < \aleph_1$.

<u>Proof</u>: Suppose $\alpha < \aleph_1$ and we are given a partition $F : [\aleph_2]^\alpha \longrightarrow 2$. Define the partition $G : [\aleph_1]^{\aleph_1} \longrightarrow 2$ as follows:

> given $f \in [\aleph_1]^{\aleph_1}$, let P_f be the α-sequence from $\aleph_2\ (=\aleph_1{}^{\aleph_1}/\mu_\omega)$ determined by the sequences in $bk_\alpha(f)$, i.e., $\forall\beta < \alpha$, $P_f(\beta) = \overline{(bk_\alpha(f))(\beta)}$. Then we define $G(f)$ to be $F(P_f)$.

Now by $\aleph_1 \longrightarrow (\aleph_1)^{\aleph_1}$, let C be a size $\aleph_1$ subset of $\aleph_1$ homogeneous for G.

Since $\aleph_1{}^{\aleph_1}/\mu_\omega = \aleph_2$, and since the constant points in $\aleph_1{}^{\aleph_1}/\mu_\omega$ (i.e., members of $\aleph_1{}^{\aleph_1}/\mu_\omega$ of the form $\bar{f}$ where f is a constant function) form a size $\aleph_1$ initial segment of $\aleph_1{}^{\aleph_1}/\mu_\omega$, the nonconstant points in $\aleph_1{}^{\aleph_1}/\mu_\omega$ form a set of size $\aleph_2$. Thus as $\bar{\bar{C}} = \aleph_1$, it is clear that the set D of nonconstant points in $C^{\aleph_1}/\mu_\omega$ is a set of size $\aleph_2$. We will show that D is homogeneous for F.

We must first show that every member of D is of the form $\bar{f}$ where f is a <u>strictly increasing</u> map from $\aleph_1$ into C. This is a consequence of the normality of μ_ω and follows from these next two general facts:

<u>Fact 1</u>: If κ is measurable and μ is a normal measure on κ, then given any f in $[\kappa]^\kappa$ and $x \subseteq f''\kappa$ such that $\mu(\{\alpha | f(\alpha) \varepsilon x\}) = 1$, the map g in $[\kappa]^\kappa$ whose range is precisely x satisfies $g = f$ almost everywhere (μ).
(Proof of Fact 1: Let $A = \{\alpha | f(\alpha) \varepsilon x\}$.
If for some $\alpha \varepsilon A$ $f(\alpha) \neq g(\alpha)$, we must have $f(\alpha) = g(\beta)$ for some $\beta < \alpha$. Thus if $\{\alpha | f(\alpha) = g(\alpha)\}$ has measure 0, the map h which sends any α to the least β such that $f(\alpha) = g(\beta)$ satisfies $\mu(\{\alpha | h(\alpha) < \alpha\}) = 1$. By the normality of μ, this would imply the existence of an $\alpha_0 < \kappa$ such that $\mu(\{\alpha | h(\alpha) = \alpha_0\}) = 1$, an obvious impossibility. Thus $\{\alpha | f(\alpha) = g(\alpha)\}$ must have measure 1. ⊠)

<u>Fact 2</u>: If κ is a measurable cardinal and μ is a normal measure on κ, then for any $Q \subseteq \kappa$, any nonconstant point of Q^κ/μ is of the form $\bar{f}$ where $f : \kappa \longrightarrow Q$ is strictly increasing. (Proof of Fact 2: Suppose h is any map from κ into Q which is not constant almost everywhere.
Let p mapping κ into κ be given by

$$p(\alpha) = \text{the least } \beta \text{ such that } h(\beta) \geq h(\alpha).$$

Certainly $p(\alpha) \leq \alpha$ for every α.
If $p(\alpha) = \alpha$ for almost every α, then there is a set A_0 of μ-measure 1 (namely $A_0 = \{\alpha | p(\alpha) = \alpha\}$) such that $\alpha < \beta$ in A_0 implies $h(\alpha) < h(\beta)$. Thus by Fact 1, if f is the κ-sequence from κ which enumerates the range of h

on A_0, then $h = f$ almost everywhere and f is strictly increasing. If, on the other hand, $p(\alpha) < \alpha$ for almost every α, the normality of μ tells us that p is constant almost everywhere. Hence by the additivity of μ, h is almost everywhere equal to a constant map, a contradiction.

$\boxtimes$)

We now know (from Fact 2) that $D = \{\bar{f} | f \in [C]^{\aleph_1}\}$, and from this it follows quite easily that $F''[D]^{\alpha} \subseteq G''[C]^{\aleph_1}$ (and hence that D is homogeneous for F). For if $p \in [D]^{\alpha}$, let us use countable choice to choose an α-sequence H of members of $[C]^{\aleph_1}$ such that for every $\beta < \alpha$, $p(\beta) = \overline{H(\beta)}$. Then by 4.2, $sf_{\alpha}(H) \in [C]^{\aleph_1}$ (check the proof of 4.2 to see that $sf_{\alpha}(H)$ is in fact in $[C]^{\aleph_1}$ rather than just $[\aleph_1]^{\aleph_1}$). By the definition of G, $F(p) = G(sf_{\alpha}(H))$, and so, as p was arbitrary, $F''[D]^{\alpha} \subseteq G''[C]^{\aleph_1}$.

$\square$

From this we have an immediate corollary:

Theorem 4.4 (Solovay): Assuming AD, $\aleph_2$ is a measurable cardinal.

Proof: Immediate from Theorems 4.3 and 3.2.

$\square$

In the proof of 4.3 we use the countable axiom of choice to pick representatives from countably many equivalence classes $\bar{f}$ in $\aleph_1^{\aleph_1}/\mu_\omega$. In fact, if we were allowed to use well-ordered choice of length $\aleph_1$, then the very same proof used in 4.3 would in fact establish $\aleph_2 \longrightarrow (\aleph_2)^\alpha$ <u>for every</u> <u>$\alpha < \aleph_2$</u> (assuming $\aleph_1 \longrightarrow (\aleph_1)^{\aleph_1}$ and $\aleph_1^{\aleph_1}/\mu_\omega = \aleph_2$). Unfortunately, well-ordered choice of length $\aleph_1$ contradicts $\aleph_1 \longrightarrow (\aleph_1)^{\aleph_1}$, and so countable choice is the most we may use. The following lemma, however, lets us pick representatives from our desired equivalence classes without using <u>any</u> choice.

<u>Lemma 4.5</u>: Let κ be a measurable cardinal and μ a normal measure. Then for any $\alpha < \kappa^+$ and f in $[\kappa]^\kappa$, $bk_\alpha(f)$ consists of representatives from the first α-many nonconstant members of the ultrapower f^κ/μ, that is, for any $\beta < \alpha$, $\overline{(bk_\alpha(f))(\beta)}$ is the βth nonconstant point in f^κ/μ.

<u>Proof:</u> We will proceed by induction on $\beta < \alpha$ to show that $\overline{(bk_\alpha(f))(\beta)}$ is the βth-least nonconstant member of the ultrapower f^κ/μ.

Since $\bar{f}$ is clearly the least nonconstant member of the ultrapower f^κ/μ, we must first show that $bk_\alpha(f)(0) = f$ almost everywhere. In the construction of $bk_\alpha(f)$, for every $\eta < \kappa$, the ordinal $f(\eta)$ is placed into the κ by κ array being built on some level less than or equal to the ηth. Let h be the map which sends η to the level on which $f(\eta)$ is placed. Then if $h(\eta) < \eta$ almost everywhere, we would have $h(\eta) = \eta_0$ for some η_0 and for almost every η. But this would imply that κ-many ordinals in f are placed on row η_0, whereas by the construction, only $\overline{\overline{\eta_0}}$-many are so placed. This contradiction then tells us that $h(\eta) = \eta$ for almost every η, and hence that for almost every η, $f(\eta)$ is placed on level η. But by our construction of $bk_\alpha(f)$, this immediately tells us that for almost every η, $f(\eta)$ is placed on level η <u>in the column associated with</u> $\underline{(bk_\alpha(f))(0)}$. Thus $f = (bk_\alpha(f))(0)$ almost everywhere.

Proceeding inductively, suppose $\overline{(bk_\alpha(f))(\beta)}$ is the βth least member of f^κ/μ for each $\beta < \gamma$. We wish to show that $\overline{(bk_\alpha(f))(\gamma)}$ is the γth least member, and for this it will suffice to prove that for any g in $[f]^\kappa$, if $g < (bk_\alpha(f))(\gamma)$ a.e., then $g = (bk_\alpha(f))(\beta)$ a.e. for some $\beta < \gamma$. So suppose $g \in [f]^\kappa$ and $g < (bk_\alpha(f))(\gamma)$ a.e. As $g \subseteq f$, a normality argument used earlier in this proof shows that for almost every $\eta < \kappa$, the ordinal $g(\eta)$ is placed on level η of the κ by κ array used in building $bk_\alpha(f)$. Let A_1 be this set of η such that $g(\eta)$ appears on level η of the $bk_\alpha(f)$ array. For any η in A_1, let $h(\eta)$ be the column of the κ by κ array for $bk_\alpha(f)$ in which $g(\eta)$ lies. Clearly $h(\eta) < \eta$ for every η in A_1, and so by the normality of μ, let η_0 be such that $h(\eta) = \eta_0$ for almost every η. If $(bk_\alpha(f))(\tau)$ is the η_0^{th} column of the array for $bk_\alpha(f)$, then clearly $g = (bk_\alpha(f))(\tau)$ a.e. But since $g < (bk_\alpha(f))(\gamma)$ a.e., we must have that $\tau < \gamma$, and so our proof is complete.

□

We now present the full picture for $\aleph_2$.

Theorem 4.6: Assume $\aleph_1 \longrightarrow (\aleph_1)^{\aleph_1}$ and $\aleph_1^{\aleph_1}/\mu_\omega = \aleph_2$. Then $\aleph_2 \longrightarrow (\aleph_2)^\alpha$ for every $\alpha < \aleph_2$.

Proof: The proof here is exactly the same as that for 4.3 except for 4.3's use of countable choice. Here is what we do in this context: we are given a member p of $[D]^\alpha$ (where now α is an arbitrary ordinal less than $\aleph_2$), and we wish to exhibit a member H of $[[C]^{\aleph_1}]^\alpha$ such that for every $\beta < \alpha$, $\overline{H(\beta)} = p(\beta)$. By using 4.5, this is easy. For since $\aleph_2$ is measurable and hence regular (by 4.4), the sup of p is some ordinal, say η, less than $\aleph_2$. By 4.5, $bk_\eta(C)$ (viewing C as a member of $[\aleph_1]^{\aleph_1}$) is a member of $[[\aleph_1]^{\aleph_1}]^\eta$ which picks representatives from the first η-many nonconstant points in $C^{\aleph_1}/\mu_\omega$. Since all of p is contained among these first η-many nonconstant points, we have our representatives from every point in p. Now we would like for each of these representatives to be in $[C]^{\aleph_1}$ rather than just $[\aleph_1]^{\aleph_1}$, but by using Fact 1 in the proof of 4.3, this can easily be arranged. We can now proceed exactly as in 4.3 to finish our proof here.

□

Theorem 4.7 (Martin-Paris): Assuming AD, $\aleph_2$ is a measurable cardinal having precisely the two normal measures μ_ω and $\mu_{\aleph_1}$.

Proof: Immediate by theorems 4.6, 3.2, and 3.5.

□

Theorem 4.8: Assuming $\aleph_1 \longrightarrow (\aleph_1)^{\aleph_1}$ and $\aleph_1^{\aleph_1}/\mu = \aleph_2$, $\aleph_3$ is cofinal with $\aleph_2$.

Proof: The bulk of our proof will be to simply show that $\aleph_2^{\aleph_1}/\mu_\omega = \aleph_3$ assuming $\aleph_1 \longrightarrow (\aleph_1)^{\aleph_1}$ and $\aleph_1^{\aleph_1}/\mu_\omega = \aleph_2$. For since $\aleph_2$ is regular (by theorem 4.4), the points in $\aleph_2^{\aleph_1}/\mu_\omega$ of the form $\overline{c_\alpha}$ (where c_α is the map from $\aleph_1$ into $\aleph_2$ constantly α) are clearly cofinal. As there are $\aleph_2$-many such points, $\aleph_2^{\aleph_1}/\mu_\omega$ equaling $\aleph_3$ tells us that $\aleph_3$ is cofinal with $\aleph_2$.

So we must show that $\aleph_2^{\aleph_1}/\mu_\omega = \aleph_3$. Let us denote by γ the order-type of the subset $[\aleph_2]^{\aleph_1}/\mu_\omega$ of $\aleph_2^{\aleph_1}/\mu_\omega$, and let us first show that γ is a cardinal. As motivation for this, consider the following proof that any ordinal λ satisfying $\lambda \longrightarrow (\lambda)^2$ is a cardinal: if g is any given map of λ 1-1 onto $\overline{\overline{\lambda}}$, let G mapping $[\lambda]^2$ into 2 be given by $G(\{\alpha,\beta\}_<) = 0$ iff $g(\alpha) < g(\beta)$. By $\lambda \longrightarrow (\lambda)^2$, let C be a subset of λ of type λ such that $\overline{\overline{G''[C]^2}} = 1$. Since λ (and hence C) can be assumed to be infinite, the fact that there are no infinite descending chains of ordinals immediately tells us that $G''[C]^2 = \{0\}$. Thus $g \restriction C$ is order-preserving and as C has order-type λ, we must have $\lambda = \overline{\overline{\lambda}}$.

Now in order to prove that γ is a cardinal using the sketch just given, we would need to have $\gamma \longrightarrow (\gamma)^2$, which unfortunately we do not. We can get by, however, using $\aleph_2 \longrightarrow (\aleph_2)^{\aleph_1+\aleph_1}$.

We begin by introducing some notation:

1) when we write a member r of $[\aleph_2]^{\aleph_1+\aleph_1}$ as (p,q), we will mean that p consists of the first $\aleph_1$-many ordinals in r and that q consists of the second $\aleph_1$-many.

2) Following the remark given immediately after the shuffling lemma, if p is a member of $[\aleph_2]^{\aleph_1}$, we will denote by p_1 the first of the two sequences in $bk_2(p)$ and by p_2 the second.

Now for our argument: suppose g is a map of γ 1-1 onto $\bar{\bar{\gamma}}$. As in our motivational sketch we will find a subset of γ of type γ on which g is order-preserving. This will show that γ is a cardinal.

Let

$$F : [\aleph_2]^{\aleph_1+\aleph_1} \longrightarrow 4 \quad \text{be given by}$$

$$F(p,q) = \begin{cases} (0,0) & \text{if} \quad g(\bar{p}_1) < g(\bar{p}_2) \quad \text{and} \quad g(\bar{p}) < g(\bar{q}) \\ (0,1) & \text{if} \quad g(\bar{p}_1) < g(\bar{p}_2) \quad \text{and} \quad g(\bar{p}) > g(\bar{q}) \\ (1,0) & \text{if} \quad g(\bar{p}_1) > g(\bar{p}_2) \quad \text{and} \quad g(\bar{p}) < g(\bar{q}) \\ (1,1) & \text{if} \quad g(\bar{p}_1) > g(\bar{p}_2) \quad \text{and} \quad g(\bar{p}) > g(\bar{q}) \;, \end{cases}$$

and by $\aleph_2 \longrightarrow (\aleph_2)^{\aleph_1+\aleph_1}$, let C be a set of cardinality $\aleph_2$ homogeneous for F. (Note that by 4.6, $\aleph_2 \longrightarrow (\aleph_2)^{\aleph_1+\aleph_1}$ is true. Strictly speaking, this implies that partitions of $[\aleph_2]^{\aleph_1+\aleph_1}$ into 2 pieces have size $\aleph_2$ homogeneous sets, but by a trivial induction on n, it also implies that for any $n < \omega$, partitions of $[\aleph_2]^{\aleph_1+\aleph_1}$ into n pieces have size $\aleph_2$ homogeneous sets.)

Our partition F just described is quite similar to the partition G of our motivation argument. For given (p,q) in $[\aleph_2]^{\aleph_1+\aleph_1}$, $\{p_1,p_2\}$ is a pair of increasing $\aleph_1$-sequences from $\aleph_2$ with $p_1 < p_2$ a.e. and $\{p,q\}$ is also a pair from $[\aleph_2]^{\aleph_1}$ with $p < q$ a.e.

Claim: $F''[C]^{\aleph_1+\aleph_1} = \{(0,0)\}$.

(Proof of claim: Suppose not. Suppose $F''[C]^{\aleph_1+\aleph_1}$ equals either $\{(0,1)\}$ or $\{(1,1)\}$. Then if q_1 denotes the first $\aleph_1$-many members of C, q_2 denotes the next $\aleph_1$-many members, q_3 the next, and, in general, q_n the nth next, then by the definition of F, $g(\bar{q}_1) > g(\bar{q}_2) > g(\bar{q}_3) > \cdots\cdots > g(\bar{q}_n) > \cdots\cdots$ would be an infinite descending chain of ordinals, a contradiction. Thus $F''[C]^{\aleph_1+\aleph_1}$ cannot equal $\{(0,1)\}$ or $\{(1,1)\}$. Suppose it equals $\{(1,0)\}$. Then given any p in $[C]^{\aleph_1}$, $g(\bar{p}_1) > g(\bar{p}_2)$. But as the function bk_1 as given by 4.2 is just the identity function, 4.5 tells us that for any r in $[C]^{\aleph_1}$, $\bar{r} = \bar{r}_1$. Thus if we let p be any member of $[C]^{\aleph_1}$, we have $g(\bar{p}_1) > g(\bar{p}_2) = g(\overline{p_{21}}) > g(\overline{p_{22}}) = g(\overline{p_{221}}) > g(\overline{p_{222}}) = \cdots$, an infinite descending chain of ordinals. This contradiction tells us that $F''[C]^{\aleph_1+\aleph_1} \neq \{(1,0)\}$, and so our claim is proved. ☒)

Let us now resume the proof of our theorem. We first wish to distinguish between two types of pairs of sequences from $[\aleph_2]^{\aleph_1}$, pairs $\{x,y\}$ where $\cup x = \cup y$ (henceforth called type ⓐ), and pairs $\{x,y\}$ where $\cup x < \cup y$ (henceforth called type ⓑ). By the shuffling lemma (plus the remark following it), given any type ⓐ pair $\{x,y\}$, there is a p in $[\aleph_2]^{\aleph_1}$ such that $\{\bar{x},\bar{y}\} = \{\bar{p}_1,\bar{p}_2\}$, and by fact 1 in the proof of 4.3, given any type ⓑ pair $\{x,y\}$, there is a (p,q) in $[\aleph_2]^{\aleph_1+\aleph_1}$ such that $\{\bar{x},\bar{y}\} = \{\bar{p},\bar{q}\}$.

Since our homogeneous set C has order-type $\aleph_2$, it is clear that the set $D = \{\bar{p} \mid p \in [C]^{\aleph_1}\}$ is of type γ. Thus we will know that γ is a cardinal as soon as we can establish the following.

Claim: g restricted to D is order-preserving.

(Proof of claim: Suppose $\{\bar{x},\bar{y}\}$ is a given pair from D, $\bar{x} < \bar{y}$. If $\{x,y\}$ is of type ⓐ, the remarks above tell us that for some p in $[C]^{\aleph_1}$, $\{\bar{x},\bar{y}\} = \{\bar{p}_1,\bar{p}_2\}$. Thus if q is any member of $[C]^{\aleph_1}$ whose inf exceeds the sup of p_1, $(p,q) \in [C]^{\aleph_1+\aleph_1}$, and so $F(p,q) = (0,0)$. By the definition of F, then, $g(\bar{x}) = g(\bar{p}_1) < g(\bar{p}_2) = g(\bar{y})$. If, alternately, $\{x,y\}$ is a type ⓑ pair, the remarks above tell us that for some (p,q) in $[C]^{\aleph_1+\aleph_1}$, $\{\bar{x},\bar{y}\} = \{\bar{p},\bar{q}\}$. As $F(p,q) = (0,0)$, the definition of F tells us that $g(\bar{x}) = g(\bar{p}) < g(\bar{q}) = g(\bar{y})$. In either case, $g(\bar{x}) < g(\bar{y})$, and our claim is proved. ⊠)

We thus know that γ, the order-type of $[\aleph_2]^{\aleph_1}/\mu_\omega$, is a cardinal. What, then, about all of $\aleph_2^{\aleph_1}/\mu_\omega$? Well by fact 2 in the proof of 4.3, if $\bar{f} \in \aleph_2{}^{\aleph_1}/\mu_\omega$, f is either a constant function or is strictly increasing, and so

$$\aleph_2{}^{\aleph_1}/\mu_\omega = [\aleph_2]^{\aleph_1}/\mu_\omega \cup \{\bar{c}_\alpha \mid \alpha < \aleph_2\}.$$

As $[\aleph_2]^{\aleph_1}/\mu_\omega$ and $\{\bar{c}_\alpha | \alpha < \aleph_2\}$ are clearly cofinal in one another, $\aleph_2{}^{\aleph_1}/\mu_\omega$ is a cardinal of cofinality $\aleph_2$.

Claim: $\aleph_2{}^{\aleph_1}/\mu_\omega = \aleph_3$.

(Proof of claim: As $\aleph_1{}^{\aleph_1}/\mu_\omega$ is a proper initial segment of $\aleph_2{}^{\aleph_1}/\mu_\omega$, we know that $\aleph_2{}^{\aleph_1}/\mu_\omega$ is a cardinal greater than $\aleph_2$. On the other hand, as any initial segment of $\aleph_2{}^{\aleph_1}/\mu_\omega$ is contained in a set of the form $\alpha^{\aleph_1}/\mu_\omega$ for some $\alpha < \aleph_2$ (this is because $\aleph_2$ is regular), and as any set of the form $\alpha^{\aleph_1}/\mu_\omega$ has cardinality at most $\aleph_2$ whenever $\alpha < \aleph_2$ (this is because $\bar{\bar{\alpha}} \leq \aleph_1$ implies $\overline{\overline{\alpha^{\aleph_1}/\mu_\omega}} \leq \overline{\overline{\aleph_1{}^{\aleph_1}/\mu_\omega}} = \aleph_2$), we must conclude that $\aleph_2{}^{\aleph_1}/\mu_\omega = \aleph_3$. ⊠)

Our theorem now follows from this claim. □

Recall that a cardinal κ is <u>Jonsson</u> if every structure of cardinality κ has a proper elementary substructure of cardinality κ. With $[\kappa]^{<\omega}$ denoting the collection of finite subsets of κ, we can state Jonssoness in terms of partition relations as follows:

> κ is Jonsson iff for every partition $F : [\kappa]^{<\omega} \longrightarrow \kappa$ there exists a subset C of κ of cardinality κ such that $F''[C]^{<\omega} \neq \kappa$.

We are now ready for our next result.

Theorem 4.9: Assuming $\aleph_1 \longrightarrow (\aleph_1)^{\aleph_1}$ and $\aleph_1^{\aleph_1}/\mu_\omega = \aleph_2$, $\aleph_3$ is a Jonsson cardinal.

Proof: Let us first show that given any partition $F : [\aleph_3]^2 \longrightarrow \aleph_1$, there exists a set D of cardinality $\aleph_3$ such that $\overline{\overline{F''[D]^2}} \leq 2$: given $F : [\aleph_3]^2 \longrightarrow \aleph_1$, let G be the partition of $[\aleph_2]^{\aleph_1+\aleph_1}$ into $\aleph_1 \times \aleph_1$-many pieces defined as follows:

$$\text{for any } (p,q) \text{ in } [\aleph_2]^{\aleph_1+\aleph_1},$$

$$G(p,q) =_{df} \langle F(\{\bar{p}_1, \bar{p}_2\}),\ F(\{\bar{p}, \bar{q}\})\rangle .$$

Now by 4.6 and 3.1, there exists a size $\aleph_2$ subset C of $\aleph_2$ such that $\overline{\overline{G''[C]^{\aleph_1+\aleph_1}}} = 1$. As C has size $\aleph_2$, the proof of 4.8 just completed tells us that $D = [C]^{\aleph_1}/\mu_\omega$ has cardinality $\aleph_3$.

Claim: $\overline{\overline{F''[D]^2}} \leq 2$.

(Proof of claim: Let α and β, ordinals below $\aleph_1$, be such that $G''[C]^{\aleph_1+\aleph_1} = \{(\alpha,\beta)\}$. Then it is easy to see that any pair $\{\bar{x},\bar{y}\}$ in $[D]^2$ such that $\{x,y\}$ is of type ⓐ satisfies $F(\{\bar{x},\bar{y}\}) = \alpha$, and any pair $\{\bar{x},\bar{y}\}$ such that $\{x,y\}$ is of type ⓑ satisfies $F(\{\bar{x},\bar{y}\}) = \beta$ (showing this will clearly establish our claim). For if $\{x,y\}$ is of type ⓐ, let (p,q) in $[C]^{\aleph_1}$ be such that $\bar{p}_1 = \bar{x}$ and $\bar{p}_2 = \bar{y}$. Then as $G(p,q) = (\alpha,\beta)$, $F(\{\bar{p}_1,\bar{p}_2\}) = F(\{\bar{x},\bar{y}\}) = \alpha$. Similarly, if $\{x,y\}$ is of type ⓑ, let (p,q) in $[C]^{\aleph_1+\aleph_1}$ be such that $\bar{p} = \bar{x}$ and $\bar{q} = \bar{y}$. Then as $G(p,q) = (\alpha,\beta)$, $F(\{\bar{p},\bar{q}\}) = F(\{\bar{x},\bar{y}\}) = \beta$. This proves the claim. ☒)

In a similar way, we can prove that for any $n < \omega$, given a partition $F : [\aleph_3]^n \longrightarrow \aleph_1$, there exists a set D of cardinality $\aleph_3$ such that $\overline{\overline{F''[D]^n}} \leq 2^{n-1}$. We would do this by defining an auxiliary partition $G_F : [\aleph_2]^{\aleph_1 \cdot n} \longrightarrow \aleph_1^{2^{n-1}}$ which analysed the action of F on each of the 2^{n-1}-different types of n-element subsets of $\aleph_3$, the type of $\{\bar{x}_1,\bar{x}_2,\ldots,\bar{x}_n\}$ being determined by whether $\{\bar{x}_i,\bar{x}_{i+1}\}$ is type ⓐ or type ⓑ for each i, $1 \leq i < n$. Once a homogeneous set C is then given for G_F, $[C]^{\aleph_1}/\mu_\omega$ would, as before, satisfy

$$\overline{\overline{F''\left[[C]^{\aleph_2}/\mu_\omega\right]^n}} \leq 2^{n-1}.$$

Now what if we are given a partition $F : [\aleph_3]^{<\omega} \longrightarrow \aleph_1$? For each n, let F_n be the partition F restricted to

$[\aleph_3]^n$. Then by our work above, if we can find a single set C of cardinality $\aleph_2$ which was homogeneous simultaneously for each of the partitions G_{F_n}, then $[C]^{\aleph_1}/\mu_\omega$ would be a set of size $\aleph_3$ such that

$$\overline{\overline{F''\left[[C]^{\aleph_1}/\mu_\omega\right]^{<\omega}}} \le \aleph_0.$$

We would thus have shown that given any partition $H : [\aleph_3]^{<\omega} \longrightarrow \aleph_1$, there exists a size $\aleph_3$ subset D of $\aleph_3$ such that $\overline{\overline{H''[D]^{<\omega}}} \le \aleph_0$, and since this property of $\aleph_3$ clearly implies that $\aleph_3$ is Jonsson, we would have our result.

So suppose we are given partitions

$$G_{F_n} : [\aleph_2]^{\aleph_1 \cdot n} \longrightarrow \aleph_1^{2^{n-1}} \text{ for each } n < \omega, \text{ and we}$$

wish to find a single subset C of $\aleph_2$ of size $\aleph_2$ which is homogeneous, simultaneously, for all of them. We proceed as follows: let the partition

$$K : [\aleph_2]^{\aleph_1 \cdot \omega} \longrightarrow {}^{\aleph_0}\left(\bigcup_{n<\omega} \aleph_1^n\right)$$

be given by

$$\left(K(p^1,p^2,p^3,p^4,p^5,\ldots,p^i,\ldots)\right)(n) =_{df} G_{F_n}(p^1,p^2,\ldots,p^n)$$

(where our writing $(p^1, p^2, p^3, \ldots, p^i, \ldots)$ for a member p of $[\aleph_2]^{\aleph_1 \cdot \omega}$ indicates that p^1 is the sequence of the first $\aleph_1$-many members of p, p^2 the second, p^3 the third, etc.). Then any set C homogeneous for K is homogeneous simultaneously for each G_{F_n}, for given n, we would have $\overline{G_{F_n}"[C]^{\aleph_1 \cdot n}} \leq \overline{K"[C]^{\aleph_1 \cdot \omega}} = 1$. Thus we would have our proof if we could show that there exists a homogeneous set for K, and by the definition of K, this would clearly follow if we could prove the partition relation

$$\aleph_2 \longrightarrow (\aleph_2)^{\aleph_1 \cdot \omega}_{2^{\aleph_1}} .$$

This we now do: suppose we are given a partition

$$H : [\aleph_2]^{\aleph_1 \cdot \omega} \longrightarrow 2^{\aleph_1} .$$

By 3.1 and 4.6 we know that $\aleph_2$ satisfies

$$\aleph_2 \longrightarrow (\aleph_2)^{\aleph_1 \cdot \omega + \aleph_1 \cdot \omega}_{\aleph_1} ,$$

and so let C be a set of size $\aleph_2$ homogeneous for the partition $L : [\aleph_2]^{\aleph_1 \cdot \omega + \aleph_1 \cdot \omega} \longrightarrow \aleph_1 + 1$ given by

$$L(p,q) =_{df} \begin{cases} \text{"the least } \alpha \text{ such that} \\ \quad (H(p))(\alpha) \neq (H(q))(\alpha) \\ \quad \text{if } H(p) \neq H(q)\text{"} \\ \\ \aleph_1 \quad \text{otherwise.} \end{cases}$$

Claim: $L''[C]^{\aleph_1 \cdot \omega + \aleph_1 \cdot \omega} = \{\aleph_1\}$.

(Proof of claim: suppose not, and let (p,q,r) be any member of $[C]^{\aleph_1 \cdot \omega + \aleph_1 \cdot \omega + \aleph_1 \cdot \omega}$. Then for some $\alpha < \aleph_1$,

$$\begin{array}{ccc} (H(p))(\alpha) & \neq & (H(q))(\alpha) \\ \not\diagdown & & \not\diagup \\ & (H(r))(\alpha). & \end{array}$$

But since each of $(H(p))(\alpha)$, $(H(q))(\alpha)$, and $(H(r))(\alpha)$ is either a "0" or a "1", this is clearly impossible. This contradiction yields our claim. ☒)

By this claim just proved, we can now argue as we did in 3.1 to show that C is homogeneous for H. As H was arbitrary, we have established $\aleph_2 \longrightarrow (\aleph_2)^{\aleph_1 \cdot \omega}_{2^{\aleph_1}}$, and this was all we needed to complete the proof of our theorem.

□

Theorem 4.10: Assuming AD, $\aleph_3$ is a Jonsson cardinal cofinal with $\aleph_2$.

Proof: This follows immediately from 2.7, 2.9, 4.8, and 4.9.

□

Chapter V

In the chapter just completed, we have seen how the existence of a certain partition property on $\aleph_1$ can imply the existence of large cardinal properties for $\aleph_2$ and $\aleph_3$. The basic ideas here can be pushed even further to yield large cardinal properties for <u>each</u> $\aleph_n$ $(n < \omega)$, but the argument itself must be refined. For one thing, the properties established for $\aleph_2$ were weaker than those of $\aleph_1$ needed to establish them, and similarly, the properties established for $\aleph_3$ were weaker than those of $\aleph_2$ needed to establish them. Thus there is no sort of possible induction evident. Furthermore, even if we did try some sort of induction by iterating ultrapowers, we would quickly run into difficulties with the axiom of choice.

These problems can be overcome by using the break function to define a very homogeneous sort of ultrapower and by then dropping all the way back to $\aleph_1$'s partition property to prove results for the $\aleph_n$. This can be done abstractly, and our main theorem is as follows:

<u>Theorem 5.1</u>: Assume κ is an uncountable cardinal satisfying $\kappa \longrightarrow (\kappa)^{\kappa}$. Then there exists an infinite sequence of cardinals $\kappa_1 (=\kappa) < \kappa_2 < \kappa_3 < \kappa_4 < \kappa_5 < \cdots\cdots$ such that

1. κ_1 and κ_2 are measurable (in fact, $\kappa_2 \longrightarrow (\kappa_2)^{\alpha}$ for each $\alpha < \aleph_1$, yet $\kappa_2 \not\longrightarrow (\kappa_2)^{\kappa_2}$),

2. κ_n for $n > 2$ is a singular Jonsson cardinal cofinal with κ_2.

If, furthermore, there exists a normal measure μ on κ such that $\kappa^\kappa/\mu = \kappa^+$, then the κ_n can be taken so that

3. $\kappa_{n+1} = \kappa_n^+$ for each n,

4. $\kappa_2 \longrightarrow (\kappa_2)^\alpha$ for each $\alpha < \kappa_2$,

5. the measures μ_n are the only normal measures on κ_1 or κ_2.

An immediate corollary of this result is the following:

Theorem 5.2: Assuming AD, $\aleph_1$ is measurable with μ_ω its only normal measure, $\aleph_2$ is measurable with μ_ω and $\mu_{\aleph_1}$ its only normal measures, and each $\aleph_n$ for $n > 2$ is a singular Jonsson cardinal of cofinality $\aleph_2$.

Proof of 5.2: Immediate from 2.7, 2.9, 5.1, and 3.5. □

The proof of 5.1 is fairly long and can best be presented via a series of definitions and lemmas. This we now do.

Definitions: (Keep in mind that throughout this chapter we are assuming $\kappa \longrightarrow (\kappa)^{\kappa}$ and dependent choice.) Our first goal is to describe the κ_n.

(a) We begin by defining κ° to be 1 and by then noting that the function bk_1 is simply the identity. We can now define, for each integer $n \geq 2$, S_n to be the range of $bk_{\kappa^{n-2}}$ on $[\kappa]^{\kappa}$.

(b) Suppose $n > 2$. Let H be any member of $[[\kappa]^{\kappa}]^{\kappa^{n-2}}$. Then H is a κ^{n-2}-sequence from $[\kappa]^{\kappa}$, and so we can introduce the following associated notation: H_0 denotes the κ^{n-3}-sequence from $[\kappa]^{\kappa}$ consisting of the first κ^{n-3} many members of H, H_1 denotes the κ^{n-3} sequence consisting of the next κ^{n-3} many members of H, and so forth. In this way, we define H_α for every $\alpha < \kappa$. Note: Given that $H \in [[\kappa]^{\kappa}]^{\kappa^{n-2}}$, $H_\alpha \in [[\kappa]^{\kappa}]^{\kappa^{n-3}}$ for each $\alpha < \kappa$ — the κ-many H_α are the successive component blocks of H. H_α is called the αth component of H.

(c) We next define by induction on $n \geq 2$ a relation $\sim_n$ on S_n as follows: $\sim_2$ is simply "= a.e. (μ)" (that is, $f \sim_2 g$ iff $\mu(\{\alpha | f(\alpha) = g(\alpha)\}) = 1$), and, if $\sim_k$ has been defined and H and G are two members of S_{k+1}, then $H \sim_{k+1} G$ iff ${}_{df}\mu(\{\alpha | H_\alpha \sim_k G_\alpha\}) = 1$.

Lemma 5.3: For each $n \geq 2$, $\sim_n$ is an equivalence relation on S_n.

Proof: Immediate by induction on n. □

Definition: Let us define by induction on $n \geq 2$ a relation $<_n$ on $S_n/\sim_n$ as follows: given $\bar{f}$ and $\bar{g}$ in $S_2/\sim_2$, $\bar{f} <_2 \bar{g}$ iff ${}_{df}\mu(\{\alpha | f(\alpha) < g(\alpha)\}) = 1$, and if $<_k$ has been defined and $\bar{H}$ and $\bar{G}$ are given members of $S_{k+1}/\sim_{k+1}$, then $\bar{H} <_{k+1} \bar{G}$ iff ${}_{df}\mu(\{\alpha | \bar{H}_\alpha <_k \bar{G}_\alpha\}) = 1$.

Lemma 5.4: For each $n \geq 2$, $<_n$ is a well-defined well-ordering of $S_n/\sim_n$.

Proof: Immediate by induction on n. □

Definition: For each $n \geq 2$, let us define κ_n to be the order-type of $\langle S_n/\sim_n, <_n \rangle$. In the future, the ordinal κ_n will be freely identified with $S_n/\sim_n$, where $<_n$ will always be the assumed ordering of $S_n/\sim_n$.

Lemma 5.5: $\kappa < \kappa_2$, and for each $n \geq 2$, $\kappa_n < \kappa_{n+1}$.

Proof: Our first step is to define by induction on $n \geq 1$ maps $+ : [[\kappa]^\kappa]^{\kappa^{n-1}} \times [[\kappa]^\kappa]^{\kappa^{n-2}} \longrightarrow [[\kappa]^\kappa]^{\kappa^{n-1}}$ and $2 : [[\kappa]^\kappa]^{\kappa^{n-2}} \longrightarrow [[\kappa]^\kappa]^{\kappa^{n-2}}$ (by convention $\kappa^\circ = 1$ and $[[\kappa]^\kappa]^{\kappa^{-1}} = \kappa$). We do this as follows: in the base step of $n = 1$, suppose we are given $f \in [\kappa]^\kappa$ and $\alpha \in \kappa$. Then as the map $h : \kappa \longrightarrow \kappa$ given by

$$h(\beta) = f(\beta) + \alpha$$

is not constant almost everywhere, fact 2 in the proof of 4.3 tells us that h is almost everywhere equal to a strictly increasing map. Furthermore, the proof of fact 2 presents us with such a strictly increasing map canonically (relative to h). We define "$f + \alpha$" to be this canonical map. We also define "2α" to be simply the order-type $\alpha \cdot 2$. Proceeding inductively, assuming our definitions are set through $n = k$, given $H \in [[\kappa]^\kappa]^{\kappa^k}$ and $h \in [[\kappa]^\kappa]^{\kappa^{k-1}}$, we define $H + h$ $(2h)$ to be the member of $[[\kappa]^\kappa]^{\kappa^k}$ $\left([[\kappa]^\kappa]^{\kappa^{k-1}}\right)$ whose β^{th} component is $H_\beta + h_\beta$ $(2h_\beta)$.

Claim: The maps "+" and "2" are well-defined as indicated.

(Proof: The only thing which needs to be checked here is that the range sets of these maps as we defined them are what they should be. Our proof proceeds by induction:

If $n = 1$, the claim is clearly true by constuction. Suppose the claim is true if $n = k$. Then the only thing to verify in the $n = k + 1$ case (given our inductive hypothesis) is that components remain distinct, i.e., that each κ-sequence in a given component of the form $H_\beta + h_\beta$ $(2h_\beta)$ is never greater (almost everywhere) than any κ-sequence from a component of the form $H_\alpha + h_\alpha$ $(2h_\alpha)$ for $\alpha > \beta$. The case $n = 2$ is special here and follows simply because $\eta_1 < \eta_2$ and $\beta_1 < \beta_2$ imply $\eta_1 + \beta_1 < \eta_2 + \beta_2$. For $n > 2$, our situation quickly reduces to the problem of showing that if f_1 and f_2 are two members of $[\kappa]^\kappa$ and κ-many distinct g satisfy $f_1 \underset{\text{a.e.}}{<} g \underset{\text{a.e.}}{<} f_2$, then for any η and δ, $\eta, \delta<\kappa$,

$f_1 + \eta <_{\text{a.e.}} f_2 + \delta$. We do this as follows: suppose $f_1 < f_2$ a.e., yet $f_1 + \eta \geq_{\text{a.e.}} f_2 + \delta$ for some η and δ. For simplicity suppose $\delta = 0$. Then there are at most η-many g such that $f_1 \underset{\text{a.e.}}{<} g \underset{\text{a.e.}}{<} f_2$, for if

$f_1 \underset{\text{a.e.}}{<} g \underset{\text{a.e.}}{<} f_2$, let, for each $\beta < \kappa$, $\ell(\beta)$ be the ordinal such that $f_1(\beta) + \ell(\beta) = g(\beta)$. Then for almost every β, $\ell(\beta)$ is between 0 and η, and hence the additivity of μ tells us that ℓ is constant almost everywhere. Thus $g = f_1 + \tau$ for some τ between 0 and η. As $\overline{\overline{\eta}} < \kappa$, we can take contrapositives to get our desired fact. (The case for the components $2h_\alpha$ is immediate by induction.) Our claim is thus proved. ⊠)

Using our maps "+" and "2", we define for $n \geq 1$ maps $\underline{\underline{+}} : \kappa_{n+1} \times \kappa_n \longrightarrow \kappa_{n+1}$ and $\underline{\underline{2}} : \kappa_n \longrightarrow \kappa_n$ as follows (recall that $\kappa_1 = \kappa$): $\overline{H} \underline{\underline{+}} \overline{h} =_{df} \overline{H + h}$ and $\underline{\underline{2}}\overline{h} =_{df} \overline{2h}$.

Claim: The maps $\underline{\underline{+}}$ and $\underline{\underline{2}}$ are well-defined as indicated.
(Proof: By the previous claim and by 4.2, the range sets of $\underline{\underline{+}}$ and $\underline{\underline{2}}$ are as indicated. That $\underline{\underline{+}}$ and $\underline{\underline{2}}$ are well-defined is immediate by induction. ☒)

Claim: For each $n \geq 1$, given any $\bar{h} \in \kappa_n$, $\bar{H} \underline{\underline{+}} \bar{h} < \underline{\underline{2}}\bar{H}$ for every $\bar{H} \in \kappa_{n+1}$.
(Proof: This is routine for $n = 1$ and generalizes immediately by induction to all n. ☒)

Claim: For any $n \geq 1$ and $\bar{H} \in \kappa_{n+1}$, the map $\bar{h} \longmapsto \bar{H} \underline{\underline{+}} \bar{h}$ from κ_n into κ_{n+1} is order-preserving.
(Proof: By induction on n, $n = 1$ being immediate and the inductive step routine. ☒)

These last two claims show that for each $n \geq 1$, κ_n can be mapped order-preservingly into a proper initial segment of κ_{n+1}. Thus $\kappa < \kappa_2$, and for each $n \geq 2$, $\kappa_n < \kappa_{n+1}$. □

Definition: Given a size κ subset C of κ, let, for each $n \geq 2$, S_n^C denote the range of $bk_{\kappa^{n-2}}$ on $[C]^{\kappa}$.

Lemma 5.6: Given any size κ subset C of κ and given any $n \geq 2$, $\langle S_n^C/{\sim_n}, <_n\rangle$ and $\langle S_n/{\sim_n}, <_n\rangle$ are order-isomorphic, i.e., $\langle S_n^C/{\sim_n}, <_n\rangle$ has order-type κ_n.

Proof: Since $S_n^C/\sim_n \subseteq S_n/\sim_n$, $S_n^C/\sim_n$ has order-type at most that of $S_n/\sim_n$. Thus to complete our proof, we need only embed $S_n/\sim_n$ order-preservingly into $S_n^C/\sim_n$, and this we do as follows: let t be an order-preserving map of κ into C. Then consider the map ϕ which, for each f in $[\kappa]^\kappa$, sends $\overline{bk_{\kappa^{n-2}}(f)}$ to $\overline{bk_{\kappa^{n-2}}(t \circ f)}$. Suppose f and g are two members of $[\kappa]^\kappa$. Then for any α less than κ, the relationship between $f(\alpha)$ and $g(\alpha)$ is precisely that between $t \circ f(\alpha)$ and $t \circ g(\alpha)$. This observation immediately implies that ϕ is both well-defined and order-preserving, and hence we have our desired embedding.

□

Definition: For each $n \geq 2$ we wish to categorize pairs of elements from $S_n/\sim_n$ according to their degree of "entwinement". We define by induction on $n \geq 2$ what it means for a pair $\{\overline{H},\overline{G}\}$ from $S_n/\sim_n$ to be i-interlaced $(0 \leq i < n-1)$ as follows: any pair from $S_2/\sim_2$ is defined to be 0-interlaced. Suppose we have completed our induction through $n = k$, and $\{\overline{H},\overline{G}\}$, $\overline{H} <_{k+1} \overline{G}$, is a given pair from $S_{k+1}/\sim_{k+1}$. Then if $\bigcup_{\alpha<\kappa}\overline{H}_\alpha < \bigcup_{\alpha<\kappa}\overline{G}_\alpha$ (keep in mind the identification between $S_n/\sim_n$ and κ_n), we say $\{\overline{H},\overline{G}\}$ is 0-interlaced. If $\bigcup_{\alpha<\kappa}\overline{H}_\alpha = \bigcup_{\alpha<\kappa}\overline{G}_\alpha$, then we define $\{\overline{H},\overline{G}\}$ to be i+1-interlaced, where i is the unique integer satisfying $\mu(\{\alpha \mid \{\overline{H}_\alpha,\overline{G}_\alpha\}$ is i-interlaced$\}) = 1$.

Lemma 5.7: For each $n \geq 2$, κ_n is a cardinal.

Proof: We begin by introducing some notation.

Notation: We define by induction on $n > 2$ what we mean, given $H \varepsilon S_n$, by the "canonical i-interlaced m-tuple $\{H^{m,i,1}, H^{m,i,2}, \ldots\ldots, H^{m,i,m}\}$ associated with H." We do this for each $H \varepsilon S_n$ and $0 < i < n - 1$ as follows: suppose $H \varepsilon S_3$ and $m < \omega$. Then for each j, $1 \leq j \leq m$, $H^{m,1,j}$ has as its αth component H_β where H_β is the unique component of H satisfying $\overline{H}_\beta = ((bk_m(\lambda\eta[\overline{H}_\eta]))_j(\alpha)$.[1] In general, given that we have our induction through $n = k$ and are given $H \varepsilon S_{k+1}$ and $m < \omega$, then, by induction on i, $0 < i < k$, $H^{m,1,j}$ has as its αth component H_β where H_β is the unique component of H satisfying $\overline{H}_\beta = ((bk_m(\lambda\eta[\overline{H}_\eta]))_j)(\alpha)$ and $H^{m,\ell+1,j}$ has as its αth component $(H_\alpha)^{m,\ell,j}$. This completes our notational definition.

It is routinely verified that for each $H \varepsilon S_n$, m, and i, $\overline{H^{m,i,1}}, \overline{H^{m,i,2}}, \ldots\ldots, \overline{H^{m,i,m}}$ is an i-interlaced m-element subset of κ_n.

We are now ready to give our proof. (It might be useful to refer back to our proof of 4.8 for motivation here.) Suppose g is a given map of κ_n 1-1 onto $\overline{\overline{\kappa_n}}$. We wish to produce a subset D of κ_n of order-type κ_n on which g is order-preserving. Let G be the partition of $[\kappa]^\kappa$ into 2^{n-1} defined as follows:

> given $f \varepsilon [\kappa]^\kappa$, let $H_{0,f}$ and $H_{1,f}$ be the two successive blocks of κ^{n-2}-many κ-sequences from κ which make up $bk_{\kappa^{n-2}\cdot 2}(f)$. Our partition G now sends f to the

---

[1] By notation, $\lambda x[\cdots x \cdots]$ denotes the function whose value at each x is "$\cdots x \cdots$". For example, $(\lambda\alpha[\alpha^2+1])(5) = 26$.

$n-1$ sequence of 0's and 1's defined as follows:

$G(f)_0 = 0$ iff $g(\bar{H}_{0,f}) < g(\bar{H}_{1,f})$

and for $0 < i < n-1$, $G(f)_i = 0$

iff $g\left(\overline{H^{2,i,1}_{0,f}}\right) < g\left(\overline{H^{2,i,2}_{0,f}}\right)$.

As $\kappa \longrightarrow (\kappa)^{\kappa}$ implies $\kappa \longrightarrow (\kappa)^{\kappa}_{m}$ for each $m < \omega$ (by a simple induction on m), let C be a size κ subset of κ such that $\overline{\overline{G"[C]^{\kappa}}} = 1$.

<u>Claim</u>: $G"[C]^{\kappa} = (0,0,0,\ldots,0)$.

(Proof: Suppose that the claim is false, and that $f \in [C]^{\kappa}$ and $0 \leq i < n-1$ are such that $G(f)_i = 1$.

Case 1: $i = 0$: In this case $g(\bar{H}_{0,f}) > g(\bar{H}_{1,f})$. Let

$f_1 =_{df} sf_{\kappa^{n-2}}(H_{1,f})$. Then by Lemma 4.5, $\overline{H_{0,f_1}} = \overline{H_{1,f}}$, and as we clearly have $f_1 \in [C]^{\kappa}$, the homogeneity of C tells us that $g(\bar{H}_{1,f}) = g(\bar{H}_{0,f_1}) > g(\bar{H}_{1,f_1})$. Continuing in this way, we get an infinite descending chain of ordinals

$g(\bar{H}_{0,f}) > g(\bar{H}_{0,f_1}) > g(\bar{H}_{0,f_2}) > g(\bar{H}_{0,f_3}) > g(\bar{H}_{0,f_4}), \ldots\ldots$ a contradiction.

Case 2: $i > 0$: In this case $g\left(\bar{H}^{2,i,1}_{0,f}\right) > g\left(\bar{H}^{2,i,2}_{0,f}\right)$. Let

$f_1 =_{df} sf_{\kappa^{n-2}}\left(H^{2,i,2}_{0,f}\right)$. Then it is routine to verify by induction on i (using Lemma 4.5 and the fact that sf_1 is the identity) that $\bar{H}_{0,f_1} = \bar{H}^{2,i,1}_{0,f_1}$. Hence by Lemma 4.5,

$\bar{H}^{2,i,2}_{0,f} = \bar{H}^{2,i,1}_{0,f_1}$, and so by the fact that $\overline{\overline{G"[C]^{\kappa}}} = 1$,

$G(\bar{H}^{2,i,1}_{0,f}) > g(\bar{H}^{2,i,1}_{0,f_1}) > g(\bar{H}^{2,i,2}_{0,f_1})$. Continuing inductively in this way, we produce an infinite descending chain of ordinals as we did in case 1. This contradiction yields the claim.

⊠)

From this claim, we can now show that g is order-preserving on $S^C_n/\sim_n$. For if $\bar{H} < \bar{K}$ is a given pair from $S^C_n/\sim_n$, i-interlaced say, we can use Lemma 4.2 to produce a member f of $[C]^\kappa$ such that $\bar{H}_{0,f} = \bar{H}$ and $\bar{H}_{1,f} = \bar{K}$ (if $i = 0$) or $\bar{H}^{2,i,1}_{0,f} = \bar{H}$ and $\bar{H}^{2,i,2}_{0,f} = \bar{K}$ (if $i > 0$). In either case, the fact that $G(f)_i = 0$ yields $g(\bar{H}) < g(\bar{K})$. As $\bar{H} < \bar{K}$ was arbitrary, we must have that g is order-preserving on $S^C_n/\sim_n$. By Lemma 5.6, $S^C_n/\sim_n$ has type κ_n, and so g mapping $S^C_n/\sim_n$ order-preservingly into κ_n must imply $\kappa_n = \overline{\overline{\kappa_n}}$.

□

<u>Lemma 5.8</u>: For each $\alpha < \aleph_1$, $\kappa_2 \longrightarrow (\kappa_2)^\alpha$.

<u>Proof</u>: Suppose $\alpha < \aleph_1$ and $F : [\kappa_2]^\alpha \longrightarrow 2$ are given. Then we define $G : [\kappa]^\kappa \longrightarrow 2$ as follows: for any $f \in [\kappa]^\kappa$, $G(f) =_{df} F(\lambda\beta[(bk_\alpha(f))(\beta)])$. By $\kappa \longrightarrow (\kappa)^\kappa$, let C be a size κ subset of κ such that $\overline{\overline{G''[C]^\kappa}} = 1$. Let $D =_{df} S^C_2/\sim_2$. Then by Lemma 5.6, D has type κ_2, and we claim that D is homogeneous for F in the sense that $\overline{\overline{F''[D]^\alpha}} = 1$. Indeed it is easy to see that

$F''[D]^{\alpha} \subseteq G''[C]^{\kappa}$: suppose $p \in [D]^{\alpha}$. Then by using countable choice, we can find a member H of $[[C]^{\kappa}]^{\alpha}$ such that for each $\beta < \alpha$, $\overline{H(\beta)} = p(\beta)$. Clearly $F(p) = G(sf_{\alpha}(H))$, and since $sf_{\alpha}(H) \in [C]^{\kappa}$, $F(p) \in G''[C]^{\kappa}$.

□

<u>Lemma 5.9</u>: Given $\beta < \kappa^{+}$ and $C \subseteq \kappa$, $\overline{\overline{C}} = \kappa$, there exists a map G from $[\kappa]^{\kappa} \times \beta$ into $[C]^{\kappa}$ such that

(1) $f \in [\kappa]^{\kappa}$ and $\eta_1 < \eta_2 < \beta$ implies $G(f,\eta_1) <_2 G(f,\eta_2)$, and

(2) for any f in $[\kappa]^{\kappa}$ and $\eta < \beta$, $\overline{G(f,\eta)}$ is the ηth member of $S_2^C/\sim_2$ larger than $\overline{f}$.

<u>Proof</u>: Let us first fix a map h of κ 1-1 onto β. We will use h to define by induction on $\eta \leq \beta$ maps $G_\eta : [\kappa]^{\kappa} \times \eta \longrightarrow [C]^{\kappa}$ such that for every $\nu < \eta$ and $f \in [\kappa]^{\kappa}$, $\overline{G_\eta(f,\nu)}$ is the νth member of $[C]^{\kappa}$ larger than $\overline{f}$, and such that $\eta_1 < \eta_2 < \beta$ implies G_{η_2} is an extension of G_{η_1}. Our desired map G will then be taken to be G_β.

Before we actually proceed with the definition of the G_η, let us first note the following: given any ordinal $\tau < \beta$ which is a limit ordinal, we have relative to h a "canonical" at-most-κ-sequence with sup τ given as follows: since h maps κ 1-1 onto β and $\tau < \beta$, let Q be $h^{-1}\tau$. We now take our "canonical" sequence to be that whose first point is h of the least member of Q, and, inductively, whose σ^{th} point is h of the least ordinal in $Q \geq \sigma$ which h sends above every point in the sequence picked so far. It is routine to verify that this "canonical" sequence is indeed cofinal in τ.

Now for the G_η: for $x \subseteq \alpha < \kappa$, let us denote by $\alpha +^C 1$ the least member of C greater than α, and by $\overset{C}{\bigcup} x$, the least member of C greater than or equal to $\bigcup x$. Given this notation, suppose $f \in [\kappa]^\kappa$ is given. Then inductively, we let $G_1(f,0) =_{df} f$; if G_τ has been defined, and τ is a successor ordinal $\sigma + 1$, then $G_{\tau+1}(f,\tau) =_{df} \lambda\nu[(G_\tau(f,\sigma))(\nu) +^C 1]$ and $G_{\tau+1} \restriction \tau = G_\tau$; if G_τ has been defined, and τ is a limit ordinal of cofinality less than κ, and $\alpha_0, \alpha_1, \ldots, \alpha_\xi, \ldots$ $\xi < \sigma < \kappa$ is the canonical cofinal sequence for τ, then we set $G_{\tau+1}(f,\tau) =_{df} \lambda\nu\left[\overset{C}{\bigcup_{\xi<\sigma}} (G_\tau(f,\alpha_\xi))(\nu)\right]$ and $G_{\tau+1} \restriction \tau =_{df} G_\tau$; if G_τ has been defined, and τ is a limit ordinal of cofinality κ, with $\alpha_0, \alpha_1, \ldots, \alpha_\xi, \ldots$ $(\xi < \kappa)$ denoting the canonical cofinal sequence for τ, we set $G_{\tau+1}(f,\tau) =_{df} \lambda\nu\left[\overset{C}{\bigcup_{\xi<\nu}} (G_\tau(f,\alpha_\xi))(\nu)\right]$ and $G_{\tau+1} \restriction \tau =_{df} G_\tau$; if G_σ has been defined for $\sigma < \tau$ where τ is a limit ordinal, then we set $G_\tau =_{df} \bigcup_{\sigma<\tau} G_\sigma$. This completes our inductive definition of the G_η.

We now take our desired map G to be G_β. It is fairly easy to check that G behaves as intended.

□

Lemma 5.10: For every $n \geq 2$, κ_n has cofinality κ_2.

Proof: By Lemma 5.9, there exists a map $K : [\kappa]^\kappa \times \kappa^{n-2} \longrightarrow [\kappa]^\kappa$ such that for every f in $[\kappa]^\kappa$ and $\beta < \kappa^{n-2}$, $\overline{K(f,\beta)}$ is the βth member of κ_2 greater than $\bar{f}$. Thus for any f in $[\kappa]^\kappa$, $\lambda\beta[K(f,\beta)]$ is a member of S_n. It is now easy to see that the map which sends any α less than κ_2 to $\overline{\lambda\beta[K(f_\alpha,\beta)]}$ (where f_α is any member of $[\kappa]^\kappa$ such that $\bar{f}_\alpha = \alpha$) is a well-defined map of κ_2 into κ_n. Let k denote this map.

It is fairly immediate that the range of k is cofinal in κ_n. For suppose $\bar{H}$ is a given member of κ_n. Then $\beta \longrightarrow \overline{H(\beta)}$ is a well-defined map of κ^{n-2} into κ_2. As κ_2 is regular (by 5.8), let α exceed the sup of the range of this map. Then $k(\alpha)$ is easily checked to exceed $\bar{H}$ as a member of κ_n. We have thus shown that k maps κ_2 cofinally into κ_n, and since it is clear by its definition that the map k is nondecreasing, the fact that κ_2 is regular tells us that $cf(\kappa_n) = \kappa_2$. □

Definition: ⓐ Given a member p of $[\kappa]^\kappa$, let $_\omega p$ be the member of $[\kappa]^\kappa$ satisfying

$$_\omega p(\alpha) = \bigcup_{n<\omega} p(\omega\cdot\alpha+n) \quad \text{for all } \alpha < \kappa.$$

ⓑ Let $Q \subseteq 2^{[\kappa]^\kappa}$ be defined as follows:

$$Q = \left\{A \subseteq [\kappa]^\kappa \mid \text{ for some } p \in [\kappa]^\kappa, \ _\omega q \in A \text{ for all } q \subseteq p\right\}.$$

(Note: Keep in mind that we freely identify increasing maps from κ into κ and the ranges of such maps.)

<u>Lemma 5.11</u>: Q is a countably additive filter on $[\kappa]^\kappa$.

<u>Proof</u>: Suppose $\{A_i \mid i < \omega\} \subseteq Q$. Let us use countable choice to pick, for each $i < \omega$, a member f_i of $[\kappa]^\kappa$ such that ${}_\omega q \in A_i$ for all $q \subseteq f_i$. By simultaneously inductively thinning out each f_i, we can construct members g_i of $[\kappa]^\kappa$ such that for all i, $g_i \subseteq f_i$, and such that for any $i < j$ and $\alpha < \kappa$, $g_i(\alpha) < g_j(\alpha) < g_{i+1}(\alpha)$. Given that the maps g_i "interlock" as described above, it is easy to see that for any $q \subseteq g_0$ there exists, for each $i < \omega$, a $q_i \subseteq g_i$ such that ${}_\omega q_i = {}_\omega q$. Thus as ${}_\omega q_i \in A_i$ for all $q_i \subseteq g_i$ and all i, ${}_\omega q \in \bigcap_{i<\omega} A_i$ for all $q \subseteq g_i$. Thus $\bigcap_{i<\omega} A_i \in Q$. We have thus shown that Q is countably additive. Since $\phi \notin Q$, and since Q is clearly closed under superset, Q is a countably additive filter. □

<u>Lemma 5.12</u>: Given any <u>countably many</u> partitions $F_n : [\kappa]^\kappa \longrightarrow 2$, there exists a single size κ subset C of κ such that C is homogeneous <u>simultaneously</u> for each F_n, i.e., such that

$$\overline{\overline{F_n''[C]^\kappa}} = 1 \quad \text{for all} \quad n.$$

Proof: Given any partition $F : [\kappa]^\kappa \longrightarrow 2$, it is fairly easy to see that the collection of sets homogeneous for F, K_F, is a member of Q. For let $G : [\kappa]^\kappa \longrightarrow 2$ be given by

$$G(p) =_{df} F({}_\omega p) \quad \text{for all} \quad p \in [\kappa]^\kappa.$$

Then as $r \in [{}_\omega q]^\kappa$ implies $r = {}_\omega s$ for some $s \in [q]^\kappa$, if q is any set homogeneous for G, ${}_\omega t$ is homogeneous for F for every $t \subseteq q$, that is, ${}_\omega t \in K_F$ for every $t \subseteq q$. Thus $K_F \in Q$.

Thus if we are given countably many partitions $F_n : [\kappa]^\kappa \longrightarrow 2$, the countable additivity of the filter Q tells us that $\bigcap_{n<\omega} K_{F_n}$ is in Q. As a result, $\bigcap_{n<\omega} K_{F_n}$ is nonempty, and clearly any member of $\bigcap_{n<\omega} K_{F_n}$ is homogeneous simultaneously for each F_n. □

Remark: The following result of Henle follows quite easily from 5.12.

Theorem (Henle): ZF + DC $\vdash$ "$\kappa \longrightarrow (\kappa)^\kappa$ implies $\kappa \longrightarrow (\kappa)^\kappa_{2^\omega}$".

Proof of Henle's theorem: Given $F : [\kappa]^\kappa \longrightarrow 2^\omega$ let, for each $n < \omega$, $F_n : [\kappa]^\kappa \longrightarrow 2$ be given by

$$F_n(p) =_{df} (F(p))(n) \quad \text{for all} \quad p \in [\kappa]^{\kappa}.$$

Then it is clear that if C is homogeneous for each F_n, C is homogeneous for F.

□

<u>Lemma 5.13</u>: For each $n \geq 2$, κ_n is Jonsson.

<u>Proof</u>: The proof of 4.9 carries a good deal of motivation for the (rather lengthy) argument we are about to give, but there is still an element missing in 4.9 which is very important here. To motivate this further feature, let us look closely at the following argument used to prove that any cardinal τ satisfying $\tau \longrightarrow (\tau)^{m+m}$ satisfies $\tau \longrightarrow (\tau)^m_\sigma$ for every $\sigma < \tau$: given $\sigma < \tau$ and a partition $F : [\tau]^m \longrightarrow \sigma$, let $G : [\tau]^{m+m} \longrightarrow 2$ be given by $G(\{\alpha_1, \ldots, \alpha_m, \alpha_{m+1}, \ldots, \alpha_{2m}\}_<) = 0$ iff $F(\{\alpha_1, \ldots, \alpha_m\}) = F(\{\alpha_{m+1}, \ldots, \alpha_{2m}\})$. Now suppose C is homogeneous for G and of cardinality τ. There are two parts to the completion of the proof. Part (1) is showing that $G"[C]^{2m} = \{0\}$, and part (2) is showing that this implies C is homogeneous for F.

(Part 1): For any m, there are τ-many non-overlapping m-element subsets of C, and since $\sigma < \tau$, we must have that F is equal on some two of them, say $F(\{\alpha_1, \ldots, \alpha_m\}) = F(\{\beta_1, \ldots, \beta_m\})$. Since $\{\alpha_1, \ldots, \alpha_m\}$ and $\{\beta_1, \ldots, \beta_m\}$ are non-overlapping,

$\{\alpha_1,\ldots,\alpha_m, \beta_1,\ldots,\beta_m\} \in [C]^{m+m}$ and $G(\{\alpha_1,\ldots,\alpha_m, \beta_1,\ldots,\beta_m\}) = 0$. As C is homogeneous for G, $G''[C]^{2m} = \{0\}$.

(Part 2): From part 1, it immediately follows that C is homogeneous for F, for if $\{\alpha_1,\ldots,\alpha_m\}$ and $\{\beta_1,\ldots,\beta_m\}$ are <u>any</u> two given m-element subsets of C, let $\{\gamma_1,\ldots,\gamma_m\}$ be an m-element subset of C which overlaps neither $\{\alpha_1,\ldots,\alpha_m\}$ nor $\{\beta_1,\ldots,\beta_m\}$. Then since $G(\{\alpha_1,\ldots,\alpha_m, \gamma_1,\ldots,\gamma_m\}) = 0$ and $G(\{\beta_1,\ldots,\beta_m, \gamma_1,\ldots,\gamma_m\}) = 0$, we have $F(\{\alpha_1,\ldots,\alpha_m\}) = F(\{\gamma_1,\ldots,\gamma_m\}) = F(\{\beta_1,\ldots,\beta_m\})$.

The idea in our context of the κ_n is to use the above trick, but always to fall back on the partition relation $\kappa \longrightarrow (\kappa)^{\kappa}$ in place of $\tau \longrightarrow (\tau)^{n+n}$. Suppose $n \geq 2$. Then in analogy to 4.9, we are about to prove that for any $\gamma < \kappa_2$ and partition

$$F : [\kappa_n]^{<\omega} \longrightarrow \gamma ,$$

there exists a size κ_n subset C of κ_n such that

$$\overline{\overline{F''[C]^{<\omega}}} \leq \aleph_0 .$$

This would, of course, imply that κ_n is Jonsson.

Suppose $\gamma < \kappa_2$ and $F: [\kappa_n]^{<\omega} \longrightarrow \gamma$ is a given partition. For simplicity, let us first look at $F \upharpoonright [\kappa_n]^2$. Using the notation introduced in the proof of Lemma 5.7, let us define $G : [\kappa]^{\kappa} \longrightarrow 2$ as follows: given $f \in [\kappa]^{\kappa}$, let

U_f, X_f, Y_f and Z_f be the four successive blocks of length κ^{n-2} which make up $bk_{\kappa^{n-2}\cdot 4}(f)$. Then $G(f) = 0$ iff

$$F(\{\bar{U}_f, \bar{X}_f\}) = F(\{\bar{Y}_f, \bar{Z}_f\}) \text{ and}$$

$$F(\{\bar{U}_f^{2,1,1}, \bar{U}_f^{2,1,2}\}) = F(\{\bar{Y}_f^{2,1,1}, \bar{Y}_f^{2,1,2}\}) \text{ and}$$

$$F(\{\bar{U}_f^{2,2,1}, \bar{U}_f^{2,2,2}\}) = F(\{\bar{Y}_f^{2,2,1}, \bar{Y}_f^{2,2,2}\}) \text{ and },\ldots, \text{ and}$$

$$F(\{\bar{U}_f^{2,n-2,1}, \bar{U}_f^{2,n-2,2}\}) = F(\{\bar{Y}_f^{2,n-2,1}, \bar{Y}_f^{2,n-2,2}\}).$$

Let C be a size κ subset of κ homogeneous for G in the sense that $\overline{G''[C]^{\kappa}} = 1$. Then following our motivational example, we will show that $S_n^C/\sim_n$ is homogeneous for F in the sense the $\overline{F''[S_n^C/\sim_n]^2} \leq n - 1$.

Part I: $G''[C]^{\kappa} = \{0\}$.

(Proof: Let us mark off $S_2^C/\sim_2$ into successive blocks each of length $\kappa^{n-2}\cdot 2$. Since κ_2 is a cardinal, there are κ_2-many such blocks, and let us denote by K_α the αth. Now by Lemma 5.9, we can associate with K_α an n-1-tuple of ordinals less than γ, namely

$t_\alpha =_{df} \langle F(\{\bar{U},\bar{X}\}), F(\{\bar{U}^{2,1,1}, \bar{U}^{2,1,2}\}), \ldots., \ldots.,$

$F(\{\bar{U}^{2,n-2,1}, \bar{U}^{2,n-2,2}\})\rangle$, where U and X are members of S_n^C such that for each $\beta < \kappa^{n-2}$, $\overline{U(\beta)} = \overline{K_\alpha(\beta)}$ and $\overline{X(\beta)} = \overline{K_\alpha(\kappa^{n-2} + \beta)}$. (It is routine to see by induction that this n-1-tuple of ordinals is independent of which U and X we choose). We thus have a map $\alpha \longrightarrow t_\alpha$ from κ_2

into γ^{n-1}, and since κ_2 is a cardinal greater than γ, there must exist $\alpha < \beta < \kappa_2$ such that $t_\alpha = t_\beta$. But then by Lemmas 5.9 and 4.2 we can put together an f in $[C]^\kappa$ such that $G(f) = 0$. Hence $G''[C]^\kappa = \{0\}$. ☒)

Part II: $S_n^C/\sim_n$ is homogeneous for F in the sense that $\overline{\overline{F''[S_n^C/\sim_n]^2}} \leq n - 1$.

(Proof: We show that for each i, $0 \leq i < n - 1$, F takes the same value on all i-interlaced pairs from $S_n^C/\sim_n$: suppose $\{\alpha_1,\alpha_2\}$ and $\{\beta_1,\beta_2\}$ are two i-interlaced pairs from $S_n^C/\sim_n$. Then by using the fact that κ_2 is regular, Lemmas 4.2 and 5.9 enable us to routinely put together two members f and g of $[C]^\kappa$ such that if $i = 0$, $\{\bar{U}_f,\bar{X}_f\} = \{\alpha_1\alpha_2\}$, $\{\bar{U}_g,\bar{X}_g\} = \{\beta_1,\beta_2\}$, and $\{\bar{Y}_f,\bar{Z}_f\} = \{\bar{Y}_g,\bar{Z}_g\}$, and if $i > 0$, $\{\bar{U}_f^{2,i,1}, \bar{U}_f^{2,i,2}\} = \{\alpha_1,\alpha_2\}$, $\{\bar{U}_g^{2,i,1}, \bar{U}_g^{2,i,2}\} = \{\beta_1,\beta_2\}$, and $\{\bar{Y}_f^{2,i,1}, \bar{Y}_f^{2,i,2}\} = \{\bar{Y}_g^{2,i,1}, \bar{Y}_g^{2,i,2}\}$. Since $G(f) = G(g)$, $F(\{\alpha_1,\alpha_2\}) = F(\{\beta_1,\beta_2\})$. ☒)

Now what about $F \restriction [\kappa_n]^3$? By an entirely similar argument, we produce a partition $G' : [\kappa]^\kappa \longrightarrow 2$ such that if C' is any set homogeneous for G' in the sense that $\overline{\overline{G'[C']^\kappa}} = 1$, then $S_n^{C'}/\sim_n$ is homogeneous for F in the sense that $\overline{\overline{F''[S_n^{C'}/\sim_n]^3}} \leq (n - 1)^2$. The main difference here is that there are $(n - 1)^2$ different types of triples from κ_n, depending upon the interlacings of the first two ordinals in the triple and of the last two ordinals in the triple. Similarly, the above argument generalizes to $F \upharpoonright [\kappa_n]^m$ for each m. Thus the problem of finding a single size κ_n subset D of κ_n such that

$$\overline{\overline{F''[D]^{<\omega}}} \leq \aleph_0$$

reduces to the problem of finding, given countably many partitions $G_n : [\kappa]^\kappa \longrightarrow 2$, a single size κ subset E of κ such that

$$\overline{\overline{G_n''[E]^\kappa}} = 1 \quad \text{for all} \quad n.$$

This last problem is immediately solved by Lemma 5.12, and so our proof is complete.

□

The following result is due to E. Bull. Our proof, however, is somewhat different from his.

Lemma 5.14: If $\kappa^\kappa/\mu = \kappa^+$, then for some $\gamma < \kappa$, $\kappa = \gamma^+$.

Proof: Suppose not, that is, suppose κ is a limit cardinal. Let f mapping κ into κ be given by

$$f(\alpha) = \alpha^+ \quad \text{for each} \quad \alpha < \kappa.$$

Then $B =_{df} \{\beta \in \kappa^\kappa/\mu \mid \beta < \bar{f}\}$ is clearly a proper initial segment of $[\kappa]^\kappa/\mu$, and so we will have our desired contradiction (and hence the theorem) as soon as we establish the following.

<u>Claim</u>: B has cardinality <u>greater</u> than κ.

(Proof: Suppose not. Then for some $\alpha < \kappa^+$, B consists of the first α-many members of the ultrapower $[\kappa]^\kappa/\mu$, and so by 4.5,

$$B \subseteq \left\{\overline{(bk_\alpha(\kappa))(\beta)} \mid \beta < \alpha\right\}.$$

By the definition of $bk_\alpha(\kappa)$, for every $\eta < \kappa$, the ηth row of the array associated with $bk_\alpha(\kappa)$ has at most $\bar{\eta}$-many different ordinals appearing in it. Furthermore, if for every $\nu < \kappa$, $k(\nu)$ denotes the row of the array ν appears in, then $k(\nu) \leq \nu$ for every ν. If $k(\nu) < \nu$ a.e., the normality of μ would yield $k(\nu) = \nu_0$ a.e., an obvious contradiction. Thus $k(\nu) = \nu$ a.e. But if $k(\nu) = \nu$, the fact that row ν has at most $\bar{\nu}$-many distinct ordinals in it implies that there is an ordinal greater than every ordinal in the νth row yet less than ν^+. Let $\ell(\nu)$ be the least such ordinal. Then clearly $\ell < f$ a.e., yet $(bk_\alpha(\kappa))(\beta) < \ell$ a.e. for every $\beta < \alpha$. This contradicts the fact that $B \subseteq \{\overline{(bk_\alpha(\kappa))(\beta)} \mid \beta < \alpha\}$, and so our claim is proved. ⊠)

□

<u>Lemma 5.15</u>: If $\kappa^\kappa/\mu = \kappa^+$, then for each $n \geq 1$, $\kappa_{n+1} = \kappa_n^+$ (thinking of κ as κ_1).

Proof: It is routine to verify by induction that for each $n \geq 1$, $S_{n+1}/\sim_{n+1}$ is a subset of the ultrapower κ_n^κ/μ. If $\kappa^\kappa/\mu = \kappa^+$, the fact that $\kappa < \kappa_2 \subseteq \kappa^\kappa/\mu$ tells us immediately that $\kappa_2 = \kappa_1^+ = \kappa_1^{\kappa_1}/\mu$. Continuing inductively, suppose we have shown $\kappa_{k+1} = \kappa_k^+ = \kappa_k^\kappa/\mu$, and we wish to show $\kappa_{k+2} = \kappa_{k+1}^+ = \kappa_{k+1}^\kappa/\mu$. Since the cofinality of κ_{k+1} is κ_2 and since $\kappa_2 > \kappa_1$, it follows that the constant points in κ_{k+1}^κ/μ are cofinal. As a result, any initial segment of κ_{k+1}^κ/μ must have cardinality $\leq \overline{\overline{\kappa_k^\kappa/\mu}} = \kappa_{k+1}$, and so $\kappa_{k+1}^\kappa/\mu \leq \kappa_{k+1}^+$. But $\kappa_{k+1} < \kappa_{k+2} \subseteq \kappa_{k+1}^\kappa/\mu$, and so $\kappa_{k+2} = \kappa_{k+1}^+ = \kappa_{k+1}^\kappa/\mu$. □

Lemma 5.16: If $\kappa^\kappa/\mu = \kappa^+$, then $\kappa_2 \longrightarrow (\kappa_2)^\alpha$ for every $\alpha < \kappa_2$.

Proof: Our proof here is almost identical to our proof of 5.8. We proceed as we did there (assuming now that α is an ordinal ordinal less than κ_2), but we replace the sentence "Then by using countable choice can find a member H of $[[C]^{\kappa}]^\alpha$ such that for each $\beta < \alpha$, $\overline{H(\beta)} = \ell(\beta)$." with the following: "We must find a member H of $[[C]^{\kappa}]^\alpha$ such that for each $\beta < \alpha$, $\overline{H(\beta)} = \ell(\beta)$. However, since α may be uncountable, countable choice does not do this for us. What we must do is appeal to Lemma 4.5: since κ_2 is regular, and since $\ell \in [\kappa_2]^\alpha$ where $\alpha < \kappa_2$, there is an $\eta < \kappa_2$ such that

$\ell(\beta) < \eta$ for all $\beta < \alpha$. By Lemma 4.5, for each $\gamma < \eta$, $(bk_\eta(C))(\gamma)$ is a representative from the equivalence class which corresponds to the γth largest nonconstant point in C^κ/μ. Thus as the range of ℓ is contained among the first η-many members of C^κ/μ, we can select the appropriate γ's so that the $(bk_\eta(C))(\gamma)$ we are left with are representatives from each point in the range of ℓ. By the definition of bk_η and by fact 1 of the proof of 4.3, we can immediately convert this special collection of $(bk_\eta(C))(\gamma)$'s into a member H of $[[C]^\kappa]^\alpha$ such that for each $\beta < \alpha$, $\overline{H(\beta)} = \ell(\beta)$."

Other than this replacement just indicated, our proof here and the proof of 5.8 are identical. □

Proof of Theorem 4.1: Since $\overline{\overline{\bigcup Q - \bigcap Q}} \leq \kappa$, there is some $\beta < \kappa^+$ and $h \in [f]^\kappa$ such that Q is contained among the first β-many members of f^κ/μ larger than h. By Lemma 5.9 and the proof of 5.16, let $h_0, h_1, h_2, \ldots\ldots, h_\eta, \ldots\ldots, (\eta < \tau)$ be members of $[f]^\kappa$ such that for each $\eta < \tau$, $\bar{h}_\eta$ is the ηth largest member of Q. Then by 4.2, there is a g in $[f]^\kappa$ such that for each $\eta < \tau$,

$$\overline{(bk_\tau(g))(\eta)} = \bar{h}_\eta .$$

By Lemma 4.5, Q is an initial segment of $[g]^\kappa/\mu$. □

Remark: To prove 4.1 with $\overline{\overline{Q}} \leq \kappa$ in place of $\overline{\overline{\bigcup Q - \bigcap Q}} \leq \kappa$, use well-ordered choice of length κ instead of Lemma 5.9 in the above proof. This is similar to our use of countable choice in proving 5.8.

Lemma 5.17: For any $p \in [\kappa]^\kappa$, $\{\overline{{}_\omega t} \mid t \in [p]^\kappa\}$ is an ω-closed unbounded subset of κ_2.

Proof: As $p \in [\kappa]^\kappa$, it is easy to see that $\{\overline{{}_\omega t} \mid t \in [p]^\kappa\}$ is unbounded in κ_2. Suppose $p_1, p_2, \ldots, p_n, \ldots$ is an ω-sequence of members of $\{{}_\omega t \mid t \in [p]^\kappa\}$. For each $i < \omega$, let $A_i = \{\alpha \mid p_i(\alpha) < p_{i+1}(\alpha)\}$. As $\mu_\omega(A_i) = 1$ for each $i < \omega$, $\mu_\omega(\bigcap_{i<\omega} A_i) = 1$. By fact 1 in the proof of 4.3, if we let p_i' be the enumeration of the range of p_i on $\bigcap_{i<\omega} A_i$, then $\overline{p_i} = \overline{p_i'}$ for all $i < \omega$. It is immediate that if q satisfies $q(\alpha) = \bigcup_{n<\omega} p_n'(\alpha)$ for all $\alpha < \kappa$, then

$$\overline{q} = \bigcup_{n<\omega} \overline{p_n'} = \bigcup_{n<\omega} \overline{p_n} .$$

We need thus only show that $q = {}_\omega t$ for some $t \in [p]^\kappa$.

Claim: For a.e. α, $\alpha \geq \bigcup_{\beta<\alpha} q(\beta)$.

(Proof: If not, the map k given by

$$k(\alpha) = \text{"least } \beta \text{ such that } q(\beta) > \alpha\text{"}$$

would satisfy

$$k(\alpha) < \alpha \quad \text{a.e.},$$

and hence, by normality, k would be constant a.e. This is clearly a contradiction. ☒)

Let $A = \{\alpha \mid \alpha \geq \bigcup_{\beta<\alpha} q(\beta)\}$, and let $r \in [\kappa]^{\kappa}$ be the enumeration of $q''A$. By fact 1 in the proof of 4.3, $r = q$ a.e. It is now fairly easy to find a t in $[p]^{\kappa}$ such that $r = {}_{\omega}t$. For since q is the sup of the ω-sequence $p_1', p_2', \ldots,$ we clearly have $q(\alpha) > \alpha$ for every $\alpha < \kappa$. Thus given $\delta < \kappa$,

$$r(\delta) = q(\alpha) > \alpha \geq \bigcup_{\beta<\alpha} q(\beta)$$

$$\parallel$$

$$\bigcup_{n<\omega} p_n'(\alpha) .$$

We can define n_0 to be the least integer such that

$$p_{n_0}'(\alpha) > \alpha,$$

and for each $i < \omega$, define $t(\omega \cdot \delta + i)$ to be the least member η of p larger than $t \upharpoonright \omega \cdot \delta + i$ such that

$$p'_{n_0+i}(\alpha) < \eta < p'_{n_0+i+1}(\alpha).$$

(Note: Since for each $i < \omega$ and $\alpha < \kappa$, $p'_{n_0+i}(\alpha) < p'_{n_0+i+1}(\alpha)$, and $p_{n_0+i+1} = {}_{\omega}k$ for some $k \in [p]^{\kappa}$, there clearly exists such an η.)

It is now routine to check that

$$r = {}_{\omega}t,$$

and as $r \sim_2 q$ and $t \in [p]^{\kappa}$, our proof is complete.

□

<u>Lemma 5.18</u> (Henle): The ultrapower ${\kappa_2}^{\kappa_2}/\mu_{\omega}$ has cofinality κ_2.

<u>Proof</u>: We first observe that for any $A \subseteq \kappa_2$,

$\mu_{\omega}(A) = 1$ iff for some $p \in [\kappa]^{\kappa}$,

$\{{}_{\omega}t \mid t \in [p]^{\kappa}\} \subseteq A$.

For suppose $A \subseteq \kappa_2$. Let $F : [\kappa]^{\kappa} \longrightarrow 2$ by

$F(p) = 0$ iff ${}_{\omega}p \in A$ for any $p \in [\kappa]^{\kappa}$.

Let C be a size κ subset of κ such that

$$\overline{F''[C]^{\kappa}} = 1.$$

Then $\{{}_{\omega}t \mid t \in [C]^{\kappa}\} \subseteq A$ or $\{{}_{\omega}t \mid t \in [C]^{\kappa}\} \subseteq A^{c}$, depending on whether $F''[C]^{\kappa} = \{0\}$ or $F''[C]^{\kappa} = \{1\}$, respectively. By the previous lemma, it is immediate that

$$\mu_{\omega}(A) = 1 \quad \text{iff} \quad \{{}_{\omega}t \mid t \in [C]^{\kappa}\} \subseteq A.$$

What we wish to do is to produce an order-preserving map of κ_2 into $\kappa_2{}^{\kappa_2}/\mu_{\omega}$. Given $u \in [\kappa]^{\kappa}$, let f_u be the following map from κ_2 into κ_2: given $\alpha < \kappa_2$,

> let $p \in [\kappa]^{\kappa}$ be such that $\bar{p} = \alpha$.
> Then $f_u(\alpha) =_{df} \overline{u \circ p}$.

<u>Claim</u>: f_u is a well-defined member of $[\kappa_2]^{\kappa_2}$.
(Proof: Immediate as $p_1 <_2 p_2$ iff $u \circ p_1 <_2 u \circ p_2$. ☒)

We can now define

$$k : \kappa_2 \longrightarrow \kappa_2{}^{\kappa_2}/\mu_{\omega}$$

as follows:

> given $\alpha < \kappa_2$, let $u \in [\kappa]^{\kappa}$ be such that $\bar{u} = \alpha$. Then $k(\alpha) =_{df} \overline{f_u}$.

<u>Claim</u>: k is a well-defined map of κ_2 order-preservingly into $\kappa_2{}^{\kappa_2}/\mu_{\omega}$.

(Proof: Suppose $u_1 = u_2$ a.e. Let A be an ω-closed unbounded subset of $\{\alpha \mid u_1(\alpha) = u_2(\alpha)\}$, and let $B = \{\overline{{}_\omega t} \mid t \in [A]^\kappa\}$. Then $B \subseteq \kappa_2$ and $\mu_\omega(B) = 1$, and it is easy to see that $f_{u_1}(\alpha) = f_{u_2}(\alpha)$ for every α in B. For if $\alpha \in B$, let $t \in [A]^\kappa$ be such that $\overline{{}_\omega t} = \alpha$. Then

$$f_{u_1}(\alpha) = \overline{u_1 \circ {}_\omega t}$$

and

$$f_{u_2}(\alpha) = \overline{u_2 \circ {}_\omega t} .$$

But as $t \in [A]^\kappa$ and as A is ω-closed, ${}_\omega t(\beta) \in A$ for every $\beta < \kappa$. Thus for every $\beta < \kappa$,

$$(u_1 \circ {}_\omega t)(\beta) = (u_2 \circ {}_\omega t)(\beta) ,$$

and so

$$f_{u_1}(\alpha) = f_{u_2}(\alpha) .$$

As $\mu_\omega(B) = 1$,

$$f_{u_1} = f_{u_2} \quad \text{a.e.}$$

Thus k is well-defined.

To see that k is order-preserving, suppose $u_1 <_2 u_2$ is given. We must show that $f_{u_1} < f_{u_2}$ a.e. Let $C = \{\alpha \mid u_1(\alpha) < u_2(\alpha)\}$. If u_1' and u_2' are the enumerations of $u_1''C$ and $u_2''C$ respectively, then fact 1 in the proof of 4.3 tells us that $u_1' = u_1$ a.e. and $u_2' = u_2$ a.e. Thus we need only show

$$f_{u_1'} < f_{u_2'} \quad \text{a.e.}$$

But this is immediate, for given $p \in [\kappa]^\kappa$, $u_1' \circ p < u_2' \circ p$ everywhere as $u_1' < u_2'$ everywhere. ☒)

To complete the proof of our lemma, we need only establish the following.

<u>Claim</u>: $k''\kappa_2$ is unbounded in $\kappa_2{}^{\kappa_2}/\mu_\omega$.

(Proof: Suppose $f : \kappa_2 \longrightarrow \kappa_2$ is given. We must find $\alpha < \kappa_2$ such that $k(\alpha) > \bar{f}$. (By the proof of theorem 4.3, we may assume that f is strictly increasing). Let $p \in [\kappa]^\kappa$ be such that

$$\{{}_\omega t \mid t \in [p]^\kappa\} \subseteq (f''\kappa_2)_\omega.$$

Let u be the enumeration of $\{{}_\omega p(\alpha + 1) \mid \alpha$ is a limit ordinal$\}$. Then it follows that

$$k(\bar{u}) > \bar{f}.$$

To see this we argue as follows: in the previous lemma, we proved that

$$\mu_\omega\left(\left\{\alpha < \kappa_2 \mid \alpha \geq \bigcup_{\beta<\alpha} f(\beta)\right\}\right) = 1.$$

Also, for almost every $\alpha < \kappa_2$, the αth member of $(f''\kappa_2)_\omega$ is α. Otherwise, for almost every α in $(f''\kappa_2)_\omega$, α is the $t(\alpha)^{th}$ member of $(f''\kappa_2)_\omega$ where $t(\alpha) < \alpha$. By normality, t would thus be constant almost everywhere, an obvious absurdity. Let B denote

$$\left\{\alpha < \kappa_2 \mid \alpha \geq \bigcup_{\beta<\alpha} f(\beta) \text{ and } \alpha \text{ is the } \alpha^{th} \text{ largest member of } (f''\kappa_2)_\omega\right\}.$$

Then $\mu_\omega(B) = 1$, and it is immediate that for every α in B, the $\alpha + 1^{st}$ member of $(f''\kappa_2)_\omega$ exceeds the $\alpha + 1^{st}$ member of $f''\kappa_2$, i.e., $f(\alpha + 1)$. It is now fairly easy to see that for every α in B,

$$f_u(\alpha) > f(\alpha).$$

For if $\alpha \in B$ and $q \in [\kappa]^\kappa$ is such that

$$\bar{q} = \alpha,$$

then

$$\begin{aligned} f_u(\bar{q}) &= \overline{u \circ q} \\ &= \text{"the } \bar{q}^{th} \text{ largest nonconstant point in the ultrapower } u^\kappa/\mu_\omega\text{"} \\ &\geq \text{"the } \bar{q} + 1^{st} \text{ largest nonconstant point in } ({}_\omega p)^\kappa/\mu_\omega\text{"} \\ &\quad (\text{as } u > {}_\omega p \text{ everywhere, } u \circ q > {}_\omega p \circ q \text{ everywhere}) \end{aligned}$$

$\geq$ "the $\bar{q}+1^{st}$ largest member of $(f''\kappa_2)_\omega$

(as $[_\omega p]^\kappa \subseteq \{_\omega t \mid t \in [p]^\kappa\} \subseteq (f''\kappa_2)_\omega$)

$> f(\bar{q}+1)$ (as $\bar{q} \in B$)

$> f(\bar{q})$ (as f is strictly increasing).

As $f_u > f$ on B, and as $\mu_\omega(B) = 1$, we have shown

$$k(\bar{u}) > \bar{f},$$

and so our claim is proved. $\boxtimes$)

$\square$

<u>Lemma 5.19</u>: $\kappa_2 \not\longrightarrow (\kappa_2)^{\kappa_2}$.

<u>Proof</u>: If $\kappa_2 \longrightarrow (\kappa_2)^{\kappa_2}$, the proof of 5.8 would show that

$$[\kappa_2]^{\kappa_2}/\mu_\omega \longrightarrow ([\kappa_2]^{\kappa_2}/\mu_\omega)^\alpha$$

for every $\alpha < \aleph_1$. Thus by 3.2,

$$[\kappa_2]^{\kappa_2}/\mu_\omega$$

would be measurable. However, by 5.18,

$$[\kappa_2]^{\kappa_2}/\mu_\omega$$

has cofinality κ_2, and by 5.5, $\kappa_2 < [\kappa_2]^{\kappa_2}/\mu_\omega$. Thus

$$[\kappa_2]^{\kappa_2}/\mu_\omega$$

would be singular, contradicting its measurability. □

Proof of Theorem 5.1: The κ_n form an increasing sequence of cardinals by 5.5 and 5.7. (1) follows from 3.2, 5.8, and 5.19, and (2) follows from 5.10 and 5.13. (3) and (4) are immediate consequences of 5.15 and 5.16 respectively, and (5) follows from 5.14 and 3.5. □

Looking back at our proof of 5.13, we see that something stronger than just the Jonssoness of κ_n was established:

Definition: Given cardinals $\gamma > \delta$, and a function h from ω into ω, let us denote by

$$\gamma \longrightarrow [\gamma]^{<\omega}_{\delta,h}$$

the partition relation for every partition $F : [\gamma]^{<\omega} \longrightarrow \delta$, there exists a size γ subset C of γ homogeneous for F in the sense that for each $n < \omega$,

$$\overline{\overline{F''[C]^n}} \leq h(n).$$

Strictly speaking, our proof of 5.13 yields as an immediate corollary the following result:

Theorem 5.20: Assuming AD,

$$\aleph_n \longrightarrow [\aleph_n]^{<\omega}_{\aleph_1, \lambda i[(n-1)^{i-1}]}$$

for every $n < \omega$.

We wish to contrast this, now, with the following result:

Theorem 5.21: Suppose that $\gamma > \delta \geq \omega$ are cardinals, and that h is a function from ω into ω which is eventually less than the function $\lambda n[r^n]$ for every $r > 1$. Then if γ satisfies

$$\gamma \longrightarrow [\gamma]^{<\omega}_{\delta,h} \quad ,$$

γ is a Ramsey cardinal, that is, it also satisfies

$$\gamma \longrightarrow (\gamma)^{<\omega}.$$

Proof: Suppose we are given $F : [\gamma]^{<\omega} \longrightarrow 2$ and wish to find a subset C of γ of cardinality γ such that for each n, $\overline{\overline{F''[C]^n}} = 1$. We begin by defining $G : [\gamma]^{<\omega} \longrightarrow \omega$ as follows:

$$G(\{\alpha_1, \ldots, \alpha_{2^i \cdot k}\}) =_{df} 2^{F(x_1)} \cdot 3^{F(x_2)} \cdot \ldots \cdot p_\ell^{F(x_\ell)} \cdot \ldots \cdot p_k^{F(x_k)}$$

where k is a prime number exceeding 2, where for each j, p_j is the jth prime number, and where x_1 is the i-tuple consisting of the least i-many members of $\{\alpha_1, \ldots, \alpha_{2^i \cdot k}\}$, x_2 is the next i-many members, etc. By $\gamma \longrightarrow [\gamma]^{<\omega}_{\delta,h}$, let D be homogeneous for G in the sense that for each n, $\overline{\overline{G''[D]^n}} \leq h(n)$.

Claim: For some $\alpha < \gamma$, $D - \alpha$ is homogeneous for $F \restriction [\gamma]^2$ in the sense that $\overline{\overline{F''[D - \alpha]^2}} = 1$.

(Proof: Since h is eventually less than $\lambda n[(\sqrt[4]{2})^n]$, let k be a prime at least equal to 3 so large that $h(4k) < (\sqrt[4]{2})^{4k}$, that is, so large that $h(4k) < 2^k$. Now if the claim is false, let $x_1^0, x_1^1, x_2^0, x_2^1, \ldots, x_\ell^0, x_\ell^1, \ldots, x_k^0, x_k^1$ be non-overlapping 2-tuples in D such that for each ℓ between 1 and k, $0 = F(x_\ell^0) \neq F(x_\ell^1) = 1$. Then it is routine to see that by forming $2k$-tuples from D using appropriate combinations of the x_ℓ^0 and x_ℓ^1 as the first $2k$ many ordinals, we can get 2^k many distinct things in the range of G on $[D]^{2^2 k}$. But this contradicts $\overline{\overline{G''[D]^{4k}}} \leq h(4k) < 2^k$. The claim is thus proved. ☒)

Using the claim, it is immediate that $\gamma \longrightarrow (\gamma)^2$ must be true, and so γ must be regular. In addition, it is routine to check that the claim remains true in a more general context:

"For each $n \geq 1$, there exists an ordinal $\alpha < \gamma$ such that $\overline{\overline{F''[D - \alpha]^n}} = 1$."

By putting these two facts together, we can now define, for each $n \geq 1$, β_n to be the least ordinal α satisfying $\overline{\overline{F''[D - \alpha]^n}} = 1$. Then if $\beta =_{df} \bigcup_{n<\omega} \beta_n$, the regularity of γ tells us that $\beta < \gamma$, and so $C =_{df} D - \beta$ is a subset of γ of size γ such that for each n, $\overline{\overline{F''[C]^n}} = 1$. This is exactly what we needed, and so our proof is complete.

□

We close this chapter by considering the question as to whether or not there exist any interesting filters on the $\aleph_n$ for $n > 2$. For example, since by assuming AD we know that each $\aleph_n$ is Jonsson, we might ask whether or not there exists a Jonsson filter on each $\aleph_n$.

Definition: Let γ be a given uncountable cardinal.

(a) A filter U on γ is uniform if every member of U has cardinality γ.

(b) A uniform filter U on γ is a <u>Jonsson filter</u> if given any partition $F : [\gamma]^{<\omega} \longrightarrow \gamma$, there exists a homogeneous set <u>in U</u>, i.e., there exists a set A in U such that $F''[A]^{<\omega} \subsetneq \gamma$.

Jonsson filters are of great interest in set theory. For example, Prikry's method of forcing for changing cofinalities, a most powerful tool, can actually be carried out starting with just a Jonsson filter. Certainly one gets better control if he uses the method with a measurable cardinal, but it turns out that by just assuming the existence of a Jonsson filter, one can get the basic techniques into operation.

<u>Definition</u>: Given κ_n as defined in the proof of 5.1, let, for each $n \geq 3$, H_n be the following collection of subsets of κ_n: for any $A \subseteq \kappa_n$, $A \in H_n$ iff for some f in $[\kappa]^\kappa$, $\overline{bk_{\kappa^{n-2}}(_\omega g)} \in A$ for every $g \in [f]^\kappa$.

<u>Theorem 5.22</u> (Henle): Given $\kappa \longrightarrow (\kappa)^\kappa$, H_n is a countably additive Jonsson filter on κ_n for each $n \geq 3$.

<u>Proof</u>: Since it is fairly clear that for any $q \in [\kappa]^\kappa$, $r \in [_\omega q]^\kappa$ implies $r = {}_\omega s$ for some $s \in [q]^\kappa$, it follows

that for each $n < \omega$, any A in H_n contains $S_n^R/\sim_n$ for some size κ subset R of κ. In fact, we can take $R = {}_\omega f$, where $f \in [\kappa]^\kappa$ is such that

$$bk_{\kappa^{n-2}}({}_\omega g) \in A \quad \text{for every} \quad g \in [f]^\kappa.$$

Thus by Lemma 5.6, H_n is uniform for every n. Also, since $A \in H_n$ implies $S_n^{\omega^f}/\sim_n \subseteq A$ for some $f \in [\kappa]^\kappa$, the proof of Lemma 5.11 tells us that H_n is a filter (countably additive since we are assuming countable choice) for each n.

Finally, to see that each H_n is in fact a Jonsson filter on κ_n, we proceed exactly as we did in proving 5.1 (in particular Lemma 5.13 of 5.1), except that in defining our auxilliary partition $G : [\kappa]^\kappa \longrightarrow 2$, we work using

$$bk_{\kappa^{n-2} \cdot 4}({}_\omega f) \quad \text{rather}$$

than

$$bk_{\kappa^{n-2} \cdot 4}(f).$$

Given, then, a size κ subset C of κ homogeneous for G in the sense that $\overline{G''[C]^\kappa} = 1$,

$$D = \left\{\overline{bk_{\kappa^{n-2}}({}_\omega g)} \mid g \in [C]^\kappa\right\}$$

is our desired set in H_n homogeneous for our original partition.

□

Chapter VI

We have remaining the problem of showing $\aleph_\omega$ Rowbottom. A new idea is needed here, for the basic methods behind 4.9 or 5.1 simply do not apply.

Let us first recall the key definition: an infinite cardinal κ is Rowbottom if given any $\lambda < \kappa$ and partition $F : [\kappa]^{<\omega} \longrightarrow \lambda$, there exists a subset C of κ of cardinality κ such that $\overline{\overline{F''[C]^{<\omega}}} \leq \omega$. (In order to rule out the pathological cases, we say, by definition, that neither $\aleph_0$ nor $\aleph_1$ is Rowbottom.)

It is clear that any Rowbottom cardinal is Jonsson. The converse turns out to be false.

If one were to take the general proof in Chapter V that each $\aleph_n$ is Jonsson and carefully interweave the constructions for each $\aleph_n$ into one another, he would be able to come up with a proof (again from AD) that $\aleph_\omega$ is Jonsson. However, he would not be close to a proof that $\aleph_\omega$ was Rowbottom, for any homogeneous set C constructed during such an interweave would have the property that $\overline{\overline{C \cap \aleph_n}} = \aleph_n$ for infinitely many $n < \omega$, and it is easy to construct a partition (using AD) $F''[\aleph_\omega]^{<\omega} \longrightarrow \aleph_2$ such that any Rowbottom homogeneous set D must satisfy $\overline{\overline{D \cap \aleph_n}} < \aleph_n$ for all but finitely many n. This is our fundamental problem.

In order to circumvent this difficulty, we plan to make strong use of the notion of "interlacing" defined during the proof of 5.1. The following might serve as motivation: in proving Lemma 5.13, we showed how to construct, given a partition

$$F : [\kappa_n]^{<\omega} \longrightarrow \gamma, \ \gamma < \kappa_2,$$

a size κ_n subset D of κ_n such that

$$\overline{\overline{F''[D]^2}} \leq n - 1.$$

The point of our argument was to show that for each i, $0 \leq i < n - 1$, there exists an ordinal $\alpha_i < \gamma$ such that for any i-interlaced pair $\{\alpha, \beta\}$ from D, $F(\{\alpha, \beta\}) = \alpha_i$. Carrying this one step further, then, if we could find a subset D^* of D such that for some i, every pair from D^* was i-interlaced, then D^* would satisfy

$$\overline{\overline{F''[D^*]^2}} = 1.$$

This idea is the starting point for our showing $\aleph_\omega$ Rowbottom.

Throughout the remainder of this chapter, let the κ_n be as in Chapter V, and let κ_ω be the limit of the κ_n. We shall ultimately show that κ_ω is Rowbottom (assuming $\kappa \longrightarrow (\kappa)^\kappa$).

Definition: For each $n \geq 2$, let $\approx_n$ be the binary relation on κ_n given by

$$\bar{H} \approx_n \bar{G} \quad \text{iff} \quad \bar{H} = \bar{G} \quad \text{or} \quad \{\bar{H}, \bar{G}\} \text{ is n-2 interlaced.}$$

Lemma 6.1: For each $n \geq 2$, $\approx_n$ is a well-defined equivalence relation on κ_n.

Proof: Immediate by induction on n. □

Lemma 6.2: For each size κ subset C of κ, each $n \geq 2$, and each $\alpha < \kappa_n$, there exists a size κ_{n-1} $\approx_n$-equivalence class contained in $S_n^C/\sim_n - \alpha$.

Proof: Our plan is to define, for each $n > 2$, a map I_n from κ_n into κ_{n-1} such that for any $\bar{H}$ and $\bar{G}$ in κ_n, $I_n(\bar{H}) = I_n(\bar{G})$ implies $\bar{H} \approx_n \bar{G}$. If we had such a map, we would have our lemma by then simply observing that κ_n being a cardinal larger than κ_{n-1} implies that I_n restricted to any size κ_n subset of κ_n must be constant on a set of size κ_{n-1}.

We define the maps I_n by induction on $n > 2$ (if $n = 2$, our lemma is trivially true) as follows: given $\bar{H} \in \kappa_3$, $I_3(\bar{H}) =_{df} \sup\{\bar{H}_\alpha \mid \alpha < \kappa\}$. Since $cf(\kappa_2) = \kappa_2 > \kappa$, and since $\bar{H}_\alpha < \kappa_2$ for each $\alpha < \kappa$, $I_3(\bar{H}) < \kappa_2$. It is routine to see that I_3 is well-defined, and that $I_3(\bar{H}) = I_3(\bar{G})$ implies $\bar{H} \approx_3 \bar{G}$. In order to define I_n for $n > 3$, we must introduce some notation and machinery: there exists a map K from S_3 into S_2 such that for any G in S_3, $\overline{K(G)}$ is $\bigcup_{\alpha<\kappa} \overline{G(\alpha)}$. Specifically, we can define K by $(K(G))(\beta) =_{df} (G(\beta))(\beta)$ for each $\beta < \kappa$. This map K we shall henceforth refer to as K_3, and we note that by induction, for every $n > 3$, there exists a map K_n from S_n into S_{n-1} given by "for any G in S_n, $K_n(G)(\alpha) = K_{n-1}(G_\alpha)$ for each $\alpha < \kappa$." One should check that for each n, $K_n(G) \varepsilon S_{n-1}$, but this is routine by induction and Lemma 4.2. We may now define $I_n : \kappa_n \longrightarrow \kappa_{n-1}$ by $I_n(\bar{H}) =_{df} \overline{K_n(H)}$. It is easy to see that each I_n is well-defined and satisfies "$I_n(\bar{H}) = I_n(\bar{G})$ implies $\bar{H} \approx_n \bar{G}$."

□

<u>Remark</u>: By Lemma 6.2, and by the motivational discussion given earlier in this chapter, we easily have the following result:

<u>Theorem 6.3</u>: For each $n > 1$, κ_n satisfies the partition relation $\kappa_n \longrightarrow (\kappa_{n-1})^{<\omega}$, that is, for every partition $F : [\kappa_n]^{<\omega} \longrightarrow 2$, there exists a size κ_{n-1} subset C of κ such that for each n, $F''[C]^n = 1$.

With essentially no extra work, 6.3 can be improved to show

$$\kappa_n \longrightarrow (\beta)^{<\alpha} \quad \text{for every}$$
$$\alpha < \aleph_1 \quad \text{and} \quad \beta < \kappa_n, \quad \text{but}$$

our main interest is in κ_ω. Here is the result:

Theorem 6.4: κ_ω is a Rowbottom cardinal.

Proof: Let $F : [\kappa_\omega]^{<\omega} \longrightarrow \gamma$, $\gamma < \kappa_\omega$, be a given partition. We wish to find a size κ_ω subset D of κ_ω such that $\overline{\overline{F''[D]^{<\omega}}} \leq \aleph_0$.

Our first step is to follow the proof of 5.13 and describe a countable collection G partitions of $[\kappa]^\kappa$ into 2 associated with F. In fact, we initially throw into G basically those partitions of $[\kappa]^\kappa$ which arose in the proof of 5.13. The only modification here is that in associating a partition of $[\kappa]^\kappa$ with, say, $F \upharpoonright [\kappa_n]^m$, we will only want to consider the action of F on m-element subsets of κ_n whose members are all n-2-interlaced with one another. Thus, for example, instead of throwing in the partition $G : [\kappa]^\kappa \longrightarrow 2$ as given in the proof of 5.13, we would throw in the partition which sends f in $[\kappa]^\kappa$ to 0 iff

$$F\left(\left\{\overline{U_f^4}, n-2,1,\ \overline{U_f^4}, n-2,2\right\}\right) = F\left(\left\{\overline{U_f^4}, n-2,3,\ \overline{U_f^4}, n-2,4\right\}\right).$$

Thus, let us throw into G all such partitions of $[\kappa]^\kappa$ associated with the restriction of F to sets of the form $[\kappa_n]^m$, where n is larger than n_0, n_0 being the least integer satisfying $\kappa_{n_0} > \gamma$. We throw at most countably many partitions into G during this step.

We must also throw "mixture partitions" into G, that is, partitions which worry about the action of F on m-element subsets of κ_ω whose members come from different S_n. For example, we would want a partition in G which considered the action of F on triples from κ_ω, where we think of the first largest element as a member of $S_k/\sim_k$, the second largest element a member of $S_k/\sim_k$ k-2-interlaced with the first largest element, and the third largest element a member of $S_\ell/\sim_\ell$ $(\ell > k)$ which does not interlace with the first two at all. Such triples we call $S_k < S_k < S_\ell$-triples, and our associated partition G to be thrown into G is simply as follows: given f in $[\kappa]^\kappa$, let H denote $bk_{\kappa^{\ell-2}}(f)$, and let K denote the least κ^{k-2}-many κ-sequences in H. Then we simply have our partition G send f to 0 iff

$$F\left(\left\{ \overline{K^{4,k-2,1}},\ \overline{K^{4,k-2,2}},\ \overline{H^{2,\ell-2,1}} \right\}\right) =$$

$$F\left(\left\{ \overline{K^{4,k-2,3}},\ \overline{K^{4,k-2,4}},\ \overline{H^{2,\ell-2,2}} \right\}\right).$$

(Keep in mind that, for example, $\overline{K}^{4,k-2,1}$ is the $\sim_k$ equivalence class of $K^{4,k-2,1}$, whereas $\overline{H}^{2,\ell-2,1}$ is a $\sim_\ell$ equivalence class).

We complete our definition of G by throwing in all such "mixture partitions", keeping in mind that we only throw in mixture partitions for those $S_{i_1} < S_{i_2} < \cdots < S_{i_m}$-tuples such that $i_1 \le i_2 \le \cdots \le i_m$ and "$i_k = i_{k+1} \Longrightarrow$ the i_k^{th} largest and i_{k+1}^{st} largest members of the tuple are

i_k-2-interlaced, and $i_k < i_{k+1} \Longrightarrow$ the i_k^{th} largest and i_{k+1}^{st} largest members of the tuple are not at all interlaced, that is, they are equivalence classes of some K and H in S_{i_k} and $S_{i_{k+1}}$, respectively, where K is an initial segment of H".

Given G as defined above, it is clear that G is at most countable, and hence by 5.12, we can find a single size κ subset C of κ which is homogeneous for each member of G.

<u>Claim</u>: For each G in G, $G''[C]^{\kappa} = \{0\}$.

(Proof: It is important to note that the argument here is different from that for "Part 1" appearing in the proof of 5.13. We proceed as follows: suppose $G \in G$, suppose, for example, G is the partition given above associated with $S_k < S_k < S_\ell$ triples. By Lemma 6.2, let D be a size κ_{k-1} equivalence class under $\approx_k$ contained in $S_k^C/\sim_k$, let α_0 be a member of D_1, let α be a member of κ_ℓ such that for some H in S_ℓ^C such that $\bar{H} = \alpha$, there is an initial segment K of H such that $\bar{K} = \alpha_0$, and let (again by 6.2) D_2 be a size $\kappa_{\ell-1}$ equivalence class under $\approx_\ell$ contained in $S_\ell^C/\sim_\ell - \alpha$. Now associated with any triple $t = \{\alpha_1 < \alpha_2 < \alpha_3\}$, α_1, $\alpha_2 \in D_1$, $\alpha_3 \in D_2$, we have an ordinal $\alpha_t < \gamma$, namely $\alpha_t = F(\{\alpha_1, \alpha_2, \alpha_3\})$. Consider the following κ_{k-1}-many triples of this sort: the first, t_0, has as its first two ordinals the first two members of D_1, and as its third ordinal the first element of D_2; the second, t_1, has as its first two members the next two members of D_1, and as its third element the next element of

D_2. Continuing inductively in this way, we define κ_{k-1}-many $S_k < S_k < S_\ell$-triples t_α. Now since $k > n_0$, $k - 1 \geq n_0$, and so $\kappa_{k-1} > \gamma$. Thus as κ_{k-1} is a cardinal, there must be $\alpha < \beta < \kappa_{k-1}$ such that $\alpha_{t_\alpha} = \alpha_{t_\beta}$. As any two members of D_1 are $(k-2)$-interlaced, and any two members of D_2 are $(\ell-2)$-interlaced, we can now use Lemma 4.2 to find an f in $[C]^\kappa$ such that, where H denotes $bk_{\kappa^{\ell-2}}(f)$ and K denotes the least κ^{k-2}-many κ-sequences in H, $\{\bar{K}^{4,k-2,1}, \bar{K}^{4,k-2,2}, \bar{H}^{2,\ell-2,1}\} = t_\alpha$ and $\{\bar{K}^{4,k-2,3}, \bar{K}^{4,k-2,4}, \bar{H}^{2,\ell-2,2}\} = t_\beta$. By the definition of G and t_α and t_β, $G(f) = 0$, and so, as C is homogeneous for G, $G''[C]^\kappa = \{0\}$. Similarly, $G^{*}{}''[C]^\kappa = \{0\}$ for any G^* in $\mathcal{G}$. ⊠)

We are now almost ready to construct our homogeneous set for F, that is, our size κ_ω subset D of κ_ω such that $\overline{\overline{F''[D]^{<\omega}}} \leq \omega$. Let us first define by induction on $n > n_0$ sets $D_n \subseteq \kappa_n$ as follows: D_{n_0+1} is any size κ_{n_0} equivalence class under $\approx_{n_0+1}$ contained in $S^C_{n_0+1}/\sim_{n_0+1}$, and, if D_k has been defined, let α_0 be the least member of D_k, let α be the least member of κ_{k+1} such that for some H in S^C_{k+1} such that $\bar{H} = \alpha$, $\bar{H}_0 = \alpha_0$, and let D_{k+1} be a size κ_k equivalence class under $\approx_{k+1}$ contained in $S^C_{k+1}/\sim_{k+1} - \alpha$. This completes our inductive definition of the D_n, and we note that for each $n > n_0$, D_n is a size κ_{n-1} subset of κ_n.

Let us now set D equal to the union of the D_n. Then clearly, D is a size κ_ω subset of κ_ω.

Claim: D is homogeneous for F, that is, $\overline{\overline{F''[D]^{<\omega}}} \leq \omega$.

(Proof: The proof here is similar to Part II in the proof of 5.13. What we show is that all tuples from D of the same "type" are sent by F to the same place. For example, suppose $\{\alpha_1, \alpha_2, \alpha_3\}$ and $\{\beta_1, \beta_2, \beta_3\}$ are two $S_k < S_k < S_\ell$-triples from D. Then α_1, α_2, β_1, and β_2 are all in D_k, and α_3 and β_3 are in D_ℓ. Let γ_1 and γ_2 be two members of D_k larger than any of α_1, α_2, β_1, or β_2, and let γ_3 be a member of D_ℓ larger than α_3 and β_3. Then by Lemma 4.2,we can find f and g in $[C]^\kappa$ such that with $H(J)$ denoting $bk_{\kappa^{\ell-2}}(f)(bk_{\kappa^{\ell-2}}(g))$, and $K(L)$ denoting the least κ^{k-2}-many κ-sequences in $H(J)$,

$$\left\{\bar{K}^{4,k-2,1}, \bar{K}^{4,k-2,2}, \bar{H}^{2,\ell-2,1}\right\} = \{\alpha_1,\alpha_2,\alpha_3\}, \text{ and}$$

$$\left\{\bar{L}^{4,k-2,1}, \bar{L}^{4,k-2,2}, \bar{J}^{2,\ell-2,1}\right\} = \{\beta_1,\beta_2,\beta_3\}, \text{ and}$$

$$\left\{\bar{K}^{4,k-2,3}, \bar{K}^{4,k-2,4}, \bar{H}^{2,\ell-2,2}\right\} = \left\{\bar{L}^{4,k-2,3}, \bar{L}^{4,k-2,4}, \bar{J}^{2,\ell-2,2}\right\}$$

$$= \{\gamma_1,\gamma_2,\gamma_3\}.$$

Since for the partition G associated with $S_k < S_k < S_\ell$ triples, $G''[C]^\kappa = \{0\}$, the definition of G tells us that

$F(\{\alpha_1,\alpha_2,\alpha_3\}) = F(\{\gamma_1,\gamma_2,\gamma_3\}) = F(\{\beta_1,\beta_2,\beta_3\})$. This argument thus handles type $S_k < S_k < S_\ell$-triples, and in a similar way, we handle all types of tuples from D. This proves the claim.

$\boxtimes$

Our result now follows immediately from this last claim.

$\square$

Theorem 6.5: Assuming AD, $\aleph_\omega$ is a Rowbottom cardinal.

Proof: Immediate from 2.7, 2.9, 5.1 and 6.4. $\square$

Remark: As we mentioned at the beginning of this chapter, if we were to take the proof of 5.1 and carefully interweave the constructions for each κ_n simultaneously, we would have a proof that κ_ω was Jonsson. If we did this in the context of Theorem 5.22, we would quite easily have the following:

Theorem 6.6: Assuming $\kappa \longrightarrow (\kappa)^\kappa$, there exists a Jonsson filter on κ_ω, namely the set H_ω given by

$$A \in H_\omega \quad \text{iff}_{df} \quad A \cap \kappa_n \in H_n \quad \text{for all but finitely many } n.$$

As $A \in H_\omega$ implies $\overline{A \cap \kappa_n} = \kappa_n$ for all but finitely many n, our remarks earlier in this chapter show that H_ω is not a Rowbottom filter.

——— o ———

We can highlight the main results in Chapters I, III, V and VI with the following single fact:

Theorem 6.7: Assume κ is an uncountable cardinal satisfying $\kappa \longrightarrow (\kappa)^\kappa$. Then

(1) κ is a measurable cardinal,

(2) there exists a measurable cardinal κ_2 greater than κ,

(3) there exist infinitely many singular Jonsson cardinals κ_n greater than κ_2,

(4) there exists a singular Rowbottom cardinal greater than the κ_n, namely $\kappa_\omega = \bigcup_{n<\omega} \kappa_n$.

If, in addition, $\kappa^\kappa/\mu = \kappa^+$ for some normal measure μ on κ, then

(5) $\kappa_2 = \kappa^+$, $\kappa_3 = \kappa_2^+$, and in general, $\kappa_{n+1} = \kappa_n^+$,

and

(6) the μ_α are the only normal measures on either κ or κ_2.

One might well ask how strong the assumptions or conclusions of this theorem are. Here are three relevant facts which we state without proof.

Fact A (Martin, Mitchell): If there exists an uncountable cardinal κ on which μ_ω is a measure (as it is assuming $\kappa \longrightarrow (\kappa)^\kappa$), then there exist inner models with arbitrarily many measurable cardinals.

Fact B (Kroonenberg): If for some normal measure μ on a measurable cardinal κ, $\kappa^\kappa/\mu = \kappa^+$, then there exist inner models with arbitrarily many measurable cardinals.

Fact C (Kunen): If there exist two measurable cardinals in a row (i.e., a κ such that κ and κ^+ are both measurable), then there exist inner models with arbitrarily many measurable cardinals.

Chapter VII

In this chapter, we examine in more detail the conflict between the axiom of choice and infinite exponent partition relations. Our work divides itself among three theorems. In the first, we assume the axiom of well-ordered choice and show how under this condition, all infinite exponent partition relations divide into two groups, those which are provably false, and those which are provably equivalent to $\omega \longrightarrow (\omega)^{\omega}$. Our second theorem shows just how strongly $\kappa \longrightarrow (\kappa)^{\kappa}$ contradicts well-ordered choice of length κ —— we prove that assuming $\kappa \longrightarrow (\kappa)^{\kappa}$, there is no uniform proof that every α less than κ fails to satisfy $\alpha \longrightarrow (\alpha)^{\alpha}$. Finally, in our third theorem, we construct a κ additive measure on $[\kappa]^{\omega}$ under which every well-ordered subset has measure 0. Not only is this last result interesting in its own right, but it leads to further work on ultrapowers of the form κ^{I}/μ where I is non-well-orderable. Throughout this chapter we work in ZF.

Definition: Given any set X, AC_X denotes that form of the axiom of choice asserting that any collection of nonempty sets indexed by X has nonempty cartesian product.

Well-ordered choice is the assertion "AC_{κ} for each cardinal κ".

Remark: The following is a classical result of Rado:

Theorem: Assuming $AC_{2^{2^\kappa}}$, $\kappa \not\longrightarrow (\omega)^\omega$.

Proof: It is routine to see that the binary relation $\sim$ on $[\kappa]^\omega$ given by $p \sim q$ iff "a final segment of p equals a final segment of q"

is an equivalence relation. Using $AC_{2^{2^\kappa}}$, let, for each p in $[\kappa]^\omega$, p^* be a representative from p's equivalence class. Then if we define $F : [\kappa]^\omega \longrightarrow 2$ by

$F(p) = 0$ iff "the least m such that $p^* \restriction \omega - m$ equals some final segment of p is even",

F can have no infinite homogeneous set. Indeed, if $p \in [\kappa]^\omega$ and $p \restriction \omega - n$ equals some final segment of p^*, then $F(p \restriction \omega - m) \neq F(p \restriction \omega - (m+1))$.

□

Let us now proceed with the main results of this chapter.

Notation: Given any ordinal α, let α_ℓ and α_f be the limit ordinal and the finite ordinal, respectively, such that $\alpha = \alpha_\ell + \alpha_f$.

Theorem 7.1: Assume well-ordered choice. Let (*) denote the following condition on the ordinals α, β and γ:

$$(*) \quad \alpha \geq \omega + \omega \quad \underline{\text{or}} \quad \beta \geq \alpha_\ell \geq \omega + \alpha_f \quad \underline{\text{or}} \quad \gamma \geq \omega.$$

Then

(a) if (*) holds, $\kappa \longrightarrow (\alpha)^\beta_\gamma$ is false for all κ, and

(b) if (*) fails, $\kappa \longrightarrow (\alpha)^\beta_\gamma$ is true for some κ iff $\omega \longrightarrow (\omega)^\omega$.

We break the proof of 7.1 into a series of lemmas, some of which are quite interesting in their own right. From now until the completion of the proof of 7.1, we shall assume well-ordered choice. In particular, whenever we are dealing with a cardinal κ in a given proof, we make the following use of AC_κ:

> we choose for every $\beta \leq \kappa$ cofinal with ω an ω-sequence from κ whose sup is β. For each $\beta < \kappa$ cofinal with ω, let x_β be the ω-sequence cofinal in β so chosen.

Lemma 7.2: For any κ, $\kappa \not\longrightarrow (\omega + \omega)^\omega$.

Proof: Let us define $F : [\kappa]^\omega \longrightarrow 2$ as follows: for any $p \in [\kappa]^\omega$, $F(p) = 0$ iff $p(n) > x_{\cup p}(n)$ for every $n < \omega$. Now suppose C is a type $\omega + \omega$ subset of κ such that

$$\overline{F''[C]^{\omega}} = 1.$$

We will derive a contradiction: since given any $q \in [\kappa]^{\omega}$ there exists a $q' \subseteq q$, $q' \in [\kappa]^{\omega}$ such that

$$F(q') = 0,$$

it is clear that we must have

$$F''[C]^{\omega} = \{0\}.$$

Let $\beta = \bigcup C$. Then as $\bigcup x_{\beta} = \beta$, there must be an $n < \omega$, say n_0, such that $x_{\beta}(n_0)$ exceeds the first ω-many members of C. If we now let p be the member of $[C]^{\omega}$ which has as its first n_0+1-many members the first n_0+1-many ordinals in C and has as its remaining ω-many members the final ω-many ordinals in C, then clearly $F(p) = 1$. As $p \in [C]^{\omega}$, this contradicts $F''[C]^{\omega} = \{0\}$. Our result follows.

□

<u>Lemma 7.3</u>: For any κ, $\kappa \not\longrightarrow (\omega + \omega)^{\omega+\omega}$.

<u>Proof</u>: Let us define $F : [\kappa]^{\omega+\omega} \longrightarrow 2$ as follows: for any $p \in [\kappa]^{\omega+\omega}$, $F(p) = 0$ iff $n_{p_1} > n_{p_2}$, where p_1 denotes the first ω-many ordinals in p, p_2 denotes the second ω-many, and n_q for any ω-sequence q from κ denotes the least n such that $x_{\bigcup q}(n) > q(0)$.

Now since given any ω-sequence q in $[\kappa]^{\omega}$, there exists a sub-ω-sequence q' of q such that $n_{q'} > n_q$, it follows quite easily that given any p in $[\kappa]^{\omega+\omega}$, there exists a $p' \subseteq p$, $p' \in [\kappa]^{\omega+\omega}$, such that $F(p) \neq F(p')$. Thus there can be no subset C of κ of type $\omega + \omega$ such that

$$\overline{\overline{F''[C]^{\omega+\omega}}} = 1.$$

⊠

<u>Lemma 7.4</u>: For any κ, $\kappa \not\longrightarrow (\omega)^{\omega}_{\omega}$.

<u>Proof</u>: Let us define $F : [\kappa]^{\omega} \longrightarrow \omega$ as follows: for any $p \in [\kappa]^{\omega}$, $F(p) =_{df} n_p$ (recall proof of 7.3 for the notation). Since given any $q \in [\kappa]^{\omega}$ there exists a $q' \subseteq q$, $q' \in [\kappa]^{\omega}$, such that $n_{q'} > n_q$, it is clear that there is no type ω subset C of κ such that

$$\overline{\overline{F''[C]^{\omega}}} = 1.$$

⊠

We are now in a position to say something about 7.1:

<u>Proof of 7.1 Part (a)</u>: Immediate from Lemmas 7.2, 7.3, 7.4 and the general fact that $\kappa \longrightarrow (\beta')^{\alpha'}_{\gamma}$ follows from

$\kappa \longrightarrow (\beta)^{\alpha}_{\gamma}$ where

$\alpha = \delta_1 + \alpha' + \delta_2$ and

$\beta = \delta_1 + \beta' + \delta_2$ for some δ_1 and δ_2. ⊠

In order to prove part (b) of 7.1 we need three additional lemmas.

<u>Lemma 7.5</u>: If there exists a κ such that $\kappa \longrightarrow (\omega)^{\omega}$, then $\omega \longrightarrow (\omega)^{\omega}$.

<u>Proof</u>: Assume that $\omega \not\longrightarrow (\omega)^{\omega}$, and assume κ is given. We will show $\kappa \not\longrightarrow (\omega)^{\omega}$: let F mapping $[\omega]^{\omega} \longrightarrow 2$ be such that for any $p \in [\omega]^{\omega}$, there exists a $p' \subseteq p$, $p' \in [\omega]^{\omega}$, such that $F(p') \neq F(p)$. We scale F up to a map

$$G : [\kappa]^{\omega} \longrightarrow 2$$

defined by $G(p) = F(Q_p)$ for all $p \in [\kappa]^{\omega}$, where for $p \in [\kappa]^{\omega}$,

$$Q_p =_{df} \left\{ n < \omega \mid p_{\cup p}(n) \leq p(n') < p_{\cup p}(n+1) \text{ for some } n' < \omega \right\}.$$

(Note: As $p_{\cup p}$ and p have the same sup, $Q_p \in [\omega]^{\omega}$ for every $p \in [\kappa]^{\omega}$).

Claim: For every $p \in [\kappa]^{\omega}$, there exists a $p' \subseteq p$, $p' \in [\kappa]^{\omega}$, such that $G(p) \neq G(p')$.

(Proof of claim: Suppose $p \in [\kappa]^{\omega}$. Let Q be an infinite subset of Q_p such that $F(Q) \neq F(Q_p)$. Let

$$p' = \left\{ p(n) \mid p_{\cup p}(n') \leq p(n) < p_{\cup p}(n' + 1) \right.$$
$$\left. \text{for some } n' \in Q \right\}.$$

Clearly $p' \in [\kappa]^{\omega}$ and $p' \subseteq p$. As $Q_{p'} = Q$, we have

$$G(p) = F(Q_p) \neq F(Q) = F(Q_{p'}) = G(p'). \quad \boxtimes \;)$$

By the claim, there can exist no size ω subset C of κ such that

$$\overline{\overline{G''[C]^{\omega}}} = 1.$$

$\square$

Lemma 7.6: $\omega \longrightarrow (\omega)^{\omega}$ implies $\omega + \omega \longrightarrow (\omega + j)^{\omega + i}_{k}$ for any $i \leq j < \omega$ and $k < \omega$.

Proof: Suppose $i \leq j < \omega$ and $k < \omega$ and that $F : [\omega + \omega]^{\omega + i} \longrightarrow k$ is a given partition. By Ramsey's theorem (finite version – see [16]), there exists an integer n so large that the partition relation

$$n \longrightarrow (j)^i_k$$

is true. Let n_0 be the least such integer, and let

$$s_0, s_1, s_2, s_3, s_4, s_5, \ldots\ldots, s_{m_0}$$

be an enumeration of the finitely many i-element subsets of $(\omega + n_0) - \omega$. Then by induction on $n \leq m_0$, we define partitions F_{n+1} and sets $D_n \subseteq \omega$ as follows:

$D_0 = \omega$ and, inductively,

$F_{n+1} : [D_n]^\omega \longrightarrow k$ by

$F_{n+1}(x) =_{df} F(x \cup sn)$, and

D_{n+1} is a size ω subset of D_n such that

$\overline{\overline{F_{n+1}"[D_{n+1}]^\omega}} = 1.$

(Note: Assuming $\omega \longrightarrow (\omega)^\omega$, it is immediate that for any $D \subseteq \omega$ and partition $F : [D]^\omega \longrightarrow 2$, there exists a size ω subset D' of D such that

$$\overline{\overline{F"[D']^\omega}} = 1.)$$

Now for each n, $0 \leq n \leq m_0$, let $g(s_n)$ be that integer (between 0 and k) such that

$$F''_{n+1}[D_{n+1}] = \{g(s_n)\}.$$

Then $g : [(\omega + n_0) - \omega]^i \longrightarrow k$, and so by Ramsey's theorem, let y be a j-element subset of $(\omega + n_0) - \omega$ such that

$$\overline{\overline{g''[y]^i}} = 1.$$

Suppose that in fact $g''[y]^i = \{z\}$. Then by our definition of the sets D_n, it is immediate that $D_{m_0+1} \cup y$ is a type $\omega + j$ subset of $\omega + \omega$ such that

$$F''\left[D_{m_0+1} \cup y\right]^{\omega+i} = \{z\}.$$

As F was arbitrary, we have proved $\omega + \omega \longrightarrow (\omega + j)_k^{\omega+i}$.

□

<u>Lemma 7.7</u>: $\omega \longrightarrow (\omega)^\omega$ implies the existence of a cardinal κ such that $\kappa \longrightarrow (\omega + \omega + j)_k^{\omega+i}$ for any $j < i < \omega$ and $k < \omega$.

<u>Proof</u>: There are certain similarities between the proof here and that in 7.6. However, rather than use Ramsey's theorem (as we did in 7.6), we must now use the following result of Erdős and Rado (see [2]):

> given any set x, ordinal δ, and integer $n < \omega$, there exists a cardinal κ so large that the partition

relation

$$\kappa \longrightarrow (\delta)^n_x$$

is true.

(Recall that $\kappa \longrightarrow (\delta)^n_x$ denotes "For all $F : [\kappa]^n \longrightarrow x$, there exists a type δ subset C of κ such that

$$\overline{\overline{F''[C]^n}} = 1."$$

The proof of this theorem of Erdös and Rado can be given without any use of the axiom of choice (see [7]).)

By the Erdös-Rado theorem above, there exists, for each i, j, and $k < \omega$, a κ satisfying

$$\kappa \longrightarrow (\omega + j)^i_{k \times 2^{2^\omega}}.$$

Thus there must be a κ such that

$$\kappa \longrightarrow (\omega + j)^i_{k \times 2^{2^\omega}}$$

for every i, j and $k < \omega$. Let κ_0 be the least such cardinal. We will show that

$$\kappa_0 \longrightarrow (\omega + \omega + j)^{\omega+i}_k$$

for every $j < i < \omega$ and $k < \omega$: suppose

$F : [\kappa_0]^{\omega+i} \longrightarrow k$ is a given partition. For any

p in $[\kappa_0 - \omega]^i$, let $G_p : [\omega]^\omega \longrightarrow k$ by

$$G_p(x) =_{df} F(x \cup p).$$

As $\omega \longrightarrow (\omega)^\omega$ (and hence $\omega \longrightarrow (\omega)^\omega_k$), there exist, for each of the G_p, sets C such that

$$\overline{\overline{G_p'' [C]^\omega}} = 1.$$

If we let z_p denote the least integer such that for some infinite $C \subseteq \omega$

$$G_p'' [C]^\omega = \{z_p\},$$

we can define $H : [\kappa_0 - \omega]^i \longrightarrow k \times 2^{2^\omega}$ by

$$H(p) = \left\langle z_p, \ \{S \varepsilon [\omega]^\omega \mid G_p'' [S]^\omega = \{z_p\}\} \right\rangle .$$

By the Erdös-Rado theorem, let D be a type $\omega + j$ subset of $[\kappa_0 - \omega]$ such that

$$\overline{\overline{H'' [D]^i}} = 1.$$

Using this last property of D, let $z < k$ and $C \subseteq \omega$ be such that

$$\{G_p(x) \mid p \ \varepsilon \ [D]^i, \ x \ \varepsilon \ [C]^\omega\} = \{z\}.$$

Clearly

$$F''[C \cup D]^{\omega+i} = \{G_p(x) \mid p \in [D]^i, x \in [C]^\omega\},$$

and so, $C \cup D$ is a type $\omega + \omega + j$ subset of κ_0 such that

$$\overline{\overline{F''[C \cup D]^{\omega+i}}} = 1.$$

As F was arbitrary, our proof is complete. □

Proof of 7.1 Part (b): Immediate from 7.5, 7.6 and 7.7. ⧅

This completes the proof of 7.1. □

It was shown in Chapter I that although the existence of μ_ω as a κ-additive nontrivial normal measure on a cardinal κ implies that almost every α less than κ has cofinality ω, there is no uniform way to show this, that is, there is no function $f : \kappa \times \omega \longrightarrow \kappa$ such that for almost every $\alpha < \kappa$, $f(\alpha,0)$, $f(\alpha,1)$, $f(\alpha,2)$,..... is a sequence cofinal in α. We now see just how far this sort of reasoning can go.

Theorem 7.8: Assume $\kappa \longrightarrow (\kappa)^{\kappa}$. Then there is no uniform proof that every α less than κ fails to satisfy $\alpha \longrightarrow (\alpha)^{\alpha}$, that is, there is no function $f : \kappa \longrightarrow \bigcup_{\alpha<\kappa} {}^{[\alpha]^{\alpha}}2$ such that for every $\alpha < \kappa$, $f(\alpha)$ is a partition of $[\alpha]^{\alpha}$ into 2 which fails to have a type α homogeneous set.

Proof: Suppose $f : \kappa \longrightarrow \bigcup_{\alpha<\kappa} {}^{[\alpha]^{\alpha}}2$ is such that for every $\alpha < \kappa$, $f(\alpha)$ is a partition of $[\alpha]^{\alpha}$ into 2 which fails to have a type α homogeneous set. We will derive a contradiction: it is easy to see by the normality of μ_{ω} that for any p in $[\kappa]^{\kappa}$, $\alpha \cap p$ has order-type α for almost every $\alpha < \kappa$ (under our identification between $[\kappa]^{\kappa}$ and the collection of size κ subsets of κ, when we write $\alpha \cap p$, we really mean $\alpha \cap p''\kappa$). This is because given $p \in [\kappa]^{\kappa}$, the function $k : \kappa \longrightarrow \kappa$ given by $k(\alpha) = \overline{\alpha \cap p}$ satisfies $k(\alpha) \leq \alpha$ for all α. If $k(\alpha) < \alpha$ a.e., the normality of μ_{ω} tells us that $k(\alpha) = \alpha_0$ for some α_0 and almost every α, an obvious absurdity. Thus $k(\alpha) = \alpha$ a.e., i.e., $\alpha = \overline{\alpha \cap p}$ for almost every α.

We may now define a partition $G : [\kappa]^{\kappa} \longrightarrow 2$ as follows:

> given $p \in [\kappa]^{\kappa}$, let α be the least ordinal less than κ such that
>
> $$\alpha = \overline{\alpha \cap p}.$$
>
> Let $G(p) =_{df} (f(\alpha))(\alpha \cap p)$.

By $\kappa \longrightarrow (\kappa)^{\kappa}$, let C be a size κ subset of κ such that

$$\overline{\overline{G''[C]^{\kappa}}} = 1.$$

Let α_0 be the least ordinal less than κ such that

$$\overline{\alpha_0 \cap C} = \alpha_0.$$

Since we are assuming that there is no type α_0 subset of α_0 homogeneous for $f(\alpha_0)$, let x_0 and x_1 be two type α_0 subsets of $\alpha_0 \cap C$ such that

$$(f(\alpha_0))(x_0) \neq (f(\alpha_0))(x_1).$$

Then $x_0 \cup (C - \alpha_0)$ and $x_1 \cup (C - \alpha_0)$ are two members of $[C]^{\kappa}$ on which the partition G differs, and this contradicts the fact that $\overline{\overline{G''[C]^{\kappa}}} = 1$. Our theorem now follows.

□

We now turn our attention to measures on non-well-orderable sets. The following two results (which will be useful for our work here) consider the question of whether or not partitions of infinite exponent can have homogeneous sets of measure 1.

Definition: Let $\kappa \underset{\lambda}{\longrightarrow}> (\kappa)^{\beta}$ denote the following strengthened version of $\kappa \longrightarrow (\kappa)^{\beta}$:

> for any partition $F : [\kappa]^{\beta} \longrightarrow 2$,
> there exists a subset C of κ
> of μ_{λ}-measure 1 such that
>
> $$\overline{\overline{F''[C]^{\beta}}} = 1.$$

Theorem 7.9: If $\beta \leq cf(\lambda)$, then $\kappa \underset{\lambda}{\longrightarrow}> (\kappa)^{\beta}$ (assuming $\kappa \longrightarrow (\kappa)^{\lambda\cdot\beta}$ and AC_{β}).

Theorem 7.10: If $\beta > cf(\lambda)$, then $\kappa \underset{\lambda}{\not\longrightarrow}> (\kappa)^{\beta}$.

Proof of 7.9: Suppose $F : [\kappa]^{\beta} \longrightarrow 2$ is a given partition. Then we can define $G : [\kappa]^{\lambda\cdot\beta} \longrightarrow 2$ by

$$G(q) =_{df} F({}_{\lambda}q) \quad \text{for all} \quad q \in [\kappa]^{\lambda\cdot\beta}.$$

(Note: ${}_{\lambda}q$ is just like ${}_{\omega}q$ but with λ in place of ω — it is the β-sequence consisting of the sups of the β-many successive λ-sequences which make up q. Note also the following: by Theorem 3.4, we may assume that λ is a regular cardinal. This we shall do for the duration of our proof.)

Let C be a size κ subset of κ such that

$$\overline{\overline{G''[C]^{\lambda\cdot\beta}}} = 1.$$

Then $(C)_\lambda$ is of μ_λ-measure 1 and so our proof will follow from

Claim: $\overline{F''[(C)_\lambda]^\beta} = 1$.

(Proof of claim: Given $p \in [(C)_\lambda]^\beta$, use AC_β to choose for each $\alpha < \beta$ an order-preserving map p_α from λ into C such that $\bigcup_{\eta<\lambda} p_\alpha(\eta) = p(\alpha)$. Since λ is regular, $cf(p(\alpha)) = \lambda$ for all $\alpha < \beta$, and so

$$\overline{\left\{ \eta \mid p_\alpha(\eta) > \bigcup_{\xi<\alpha} p(\xi) \right\}} = \lambda$$

for each $\alpha < \beta$. Thus one can easily find a sub λ-sequence q_α of p_α (for each $\alpha < \beta$) such that the sequence q given by

$$q((\eta,\alpha)) =_{df} q_\alpha(\eta)$$

is a member of $[C]^{\lambda\cdot\beta}$. It is now immediate that $F(p) = G(q)$, and so $\overline{F''[(C)_\lambda]^\beta} = \overline{G''[C]^{\lambda\cdot\beta}} = 1$. ⊠)

□

Proof of 7.10: Let us define $F : [\kappa]^\beta \longrightarrow 2$ by

$$F(p) = 0 \quad \text{iff} \quad \bigcup p''\, cf(\lambda) = p(cf(\lambda)) \quad \text{for all} \quad p \in [\kappa]^\beta.$$

Then if C were a $cf(\lambda)$-closed unbounded subset of κ, it is immediate that the value of F on the first β-many members of C is 0, yet the value of F on the first β-many members of C less its $cf(\lambda)^{th}$ member is 1. Thus there can be no $cf(\lambda)$-closed unbounded set homogeneous for F, and so by 3.4 and the definition of μ_λ, $\kappa \underset{\hat{\lambda}}{\not\longrightarrow} > (\kappa)^\beta$.

□

——————— o ———————

Now consider the following function $\hat{\mu}$ from $[\kappa]^\omega$ into 2: for any $A \subseteq [\kappa]^\omega$,

$\hat{\mu}(A) = 1$ iff there exists a
μ_ω-measure 1 subset C of
κ such that $[C]^\omega \subseteq A$.

<u>Theorem 7.11</u>: Assuming $\kappa \longrightarrow (\kappa)^{\omega\cdot\omega\cdot 2}$ and AC_ω, $\hat{\mu}$ is a κ-additive measure on $[\kappa]^\omega$.

<u>Proof</u>: Let $\hat{E}$ be

$$\left\{A \subseteq [\kappa]^\omega \mid \hat{\mu}(A) = 1\right\}.$$

We must show that $\hat{E}$ is a κ-complete ultrafilter on $[\kappa]^\omega$. That $\hat{E}$ is an ultrafilter is immediate from 7.9. For if

$A \subseteq [\kappa]^{\omega}$, let C be a set of μ_{ω}-measure 1 homogeneous for the partition $F : [\kappa]^{\omega} \longrightarrow 2$ given by

$$F(p) = 0 \quad \text{iff}_{df} \quad p \in A.$$

Then $[C]^{\omega} \subseteq A$ or $[C]^{\omega} \subseteq A^{c}$ depending on whether $F''[C]^{\omega} = \{0\}$ or $F''[C]^{\omega} = \{1\}$, respectively. Now suppose $\{Q_{\alpha} \mid \alpha < \gamma < \kappa\} \subseteq \hat{E}$. We must show $\bigcap_{\alpha<\gamma} Q_{\alpha} \in \hat{E}$, and so consider the partition $G : [\kappa]^{\omega} \longrightarrow \gamma$ given by

$$G(p) = \begin{cases} 0 & \text{if} \quad p \in \bigcap_{\alpha<\gamma} Q_{\alpha} \\ \\ \alpha + 1 & \text{if} \quad p \notin \bigcap_{\alpha<\gamma} Q_{\alpha} \quad \text{and} \quad \alpha \quad \text{is} \\ & \text{the least} \quad \beta < \gamma \quad \text{such that} \quad p \notin Q_{\beta}. \end{cases}$$

By Lemma 3.1, $\kappa \longrightarrow (\kappa)^{\omega \cdot \omega}_{\gamma}$, and so by the proof of 7.9, there exists a subset D of κ of μ_{ω}-measure 1 such that

$$\overline{\overline{G''[D]^{\omega}}} = 1.$$

<u>Claim</u>: $G''[D]^{\omega} = \{0\}$.

(Proof of claim: Suppose not, suppose $G''[D]^{\omega} = \{\alpha + 1\}$ for some $\alpha < \gamma$. Then $p \notin Q_{\alpha}$ for every $p \in [D]^{\omega}$, i.e., $[D]^{\omega} \cap Q_{\alpha} = \phi$. But as $Q_{\alpha} \in \hat{E}$, there exists a subset D^{*} of κ of μ_{ω}-measure 1 such that

$$[D^*]^{\omega} \subseteq Q_{\alpha},$$

and so we would have

$$[D]^{\omega} \cap [D^*]^{\omega} = \phi.$$

Since D and D^* both have μ_{ω}-measure 1, $\overline{\overline{D \cap D^*}} = \kappa$, and so

$$[D]^{\omega} \cap [D^*]^{\omega} \neq \phi.$$

This contradiction yields our claim. $\boxtimes$)

From the claim and from the definition of G,

$$[D]^{\omega} \subseteq \bigcap_{\alpha<\gamma} Q_{\alpha},$$

and so

$$\bigcap_{\alpha<\gamma} Q_{\alpha} \in \hat{E}.$$

Thus $\hat{E}$ is κ-complete. $\square$

The semi-classical theorem of Rado given at the beginning of this chapter showed just how badly infinite exponent partition relations on κ fail assuming the existence of a well-ordering of $[\kappa]^{\omega}$. Turning this around we can ask just how non-well-

-orderable $[\kappa]^\omega$ is assuming infinite exponent partition relations.

Theorem 7.12: Assuming $\kappa \longrightarrow (\kappa)^{\omega\cdot\omega\cdot 2}$ and AC_ω, there exists a measure on $[\kappa]^\omega$ under which any well-ordered subset of $[\kappa]^\omega$ has measure 0.

Proof: Consider the measure $\hat{\mu}$ of 7.11. If some set of $\hat{\mu}$-measure 1 can be well-ordered, let h be a 1-1 map of some ordinal η into $[\kappa]^\omega$ such that

$$\hat{\mu}(h''\eta) = 1.$$

Then if C is a set of μ_ω-measure 1 such that

$$[C]^\omega \subseteq h''\eta,$$

and if k is any (canonical) bijection from C onto κ, the well-ordering h induces on $[C]^\omega$ can be translated via k to a well-ordering of all of $[\kappa]^\omega$. But the existence of a well-ordering of $[\kappa]^\omega$ contradicts $\kappa \longrightarrow (\kappa)^{\omega\cdot\omega\cdot 2}$, either by Rado's theorem or by the proof of 7.1.

□

REFERENCES

[1] Bull, E., to appear.

[2] Erdös, P. and Rado, R., A partition calculus in set theory, Bull. Amer. Math. Soc. 62, 427-489.

[3] Henle, J. M., to appear.

[4] Kleinberg, E. M., Strong partition properties for infinite cardinals, The Journal of Symbolic Logic, vol. 35 (1970).

[5] Kleinberg, E. M., Weak partition properties for infinite cardinals I, Proc. Amer. Math., vol. 30 (1971).

[6] Kleinberg, E. M., Infinitary combinatorics, Proceedings of the Cambridge Summer School in Mathematical Logic 1971, Lecture Notes in Mathematics #337, Springer (1973).

[7] Kleinberg, E. M. and Seiferas J. I., On infinite exponent partition relations and well-ordered choice, The Journal of Symbolic Logic, vol. 38 (1973).

[8] Kleinberg, E. M., AD $\vdash$ "The $\aleph_n$ are Jonsson cardinals and $\aleph_\omega$ is a Rowbottom cardinal", Annals of Mathematical Logic, to appear.

[9] Kleinberg, E. M., The equiconsistency of two large cardinal axioms, Fundamenta Mathematicae, to appear.

[10] Kleinberg, E. M., An example in axiomatic set theory, Advances in Mathematics, to appear.

[11] Kroonenberg, N., to appear.

[12] Kunen, K., to appear.

[13] Martin, D. A., to appear.

[14] Martin, D. A. and Paris, J. B., to appear.

[15] Mycielski, J. and Swierczkowski, S., On the Lebesgue measurability and the axiom of determinateness, Fundamenta Mathematicae, vol. 54.

[16] Ramsey, F. P., On a problem of formal logic, Proc. London Math. Soc., vol. 30 (1930).

[17] Solovay, R. M., to appear.

[18] Tinkelman, R. to appear.